SURFACES AND DISORDER

J. W. Halley

SURFACES AND DISORDER

Proceedings of the 12th Midwest Solid State
Theory Symposium held in St. Paul,
Minnesota in September, 1984

Edited by

J. W. Halley

School of Physics and Astronomy
University of Minnesota
Minneapolis, Minnesota 55455

TRANS TECH PUBLICATIONS 1985
Switzerland - Germany - UK - USA

Copyright © 1985 Trans Tech Publications Ltd, Switzerland
ISBN 0-87849-532-0

Distributed in North America by

Trans Tech Publications c/o
Karl Distributors
16 Bear Skin Neck
Rockport, MA 01966
USA

and worldwide by

Trans Tech Publications Ltd
Trans Tech House
4711 Aedermannsdorf
Switzerland

Printed in the Netherlands

TABLE OF CONTENTS

Structure of Thin Physisorbed Layers
   L.W. Bruch   1

Metals in Intimate Contact
   J. R. Smith and J. Ferrante   21

Chemical Trends of Schottky Barriers
   J. D. Dow, O. F. Sankey and R. E. Allen   39

Phase Transitions in the Laplacian Roughening Model
   D. Bruce   51

Interfacial Phase Diagrams and Equilibrium Crystal Shapes
   M. Wortis   71

The Selection Principle of Dendritic Solidification,
   E. Ben-Jacob   73

Crystal Defects in Curved Three Dimensional Space
   J. Straley   93

Two Dimensional Electron Glass,
   R. K. Kalia and P. V. Vashishta   99

Phase Transitions in Crystalline Polymers,
   P. L. Taylor   105

The Elastic Properties of Random Networks,
   M. F. Thorpe   125

Some Remarks on Diffusion on Fractals,
   H. Nakanishi   129

Electronic Wave Functions in Disordered Systems,
   C. M. Soukoulis and E. N. Economou   145

Defect Simulation and Supercomputers,
   A. B. Kunz   155

Quantum Simulations of Small Electron-Hole Complexes,
  M. A. Lee, R. Kalia and P. D. Vashista   165

Theory of Superconducting Arrays in a Magnetic Field,
  D. Stroud and W. Y. Shih   177

Electronic Structure of Alloy Semiconductors,
  L. C. Davis   197

PREFACE

The 12th annual Midwest Solid State Theory Symposium held in St. Paul, Minnesota in 1984 attracted a very high quality of both young and established theorists from academic and industrial institutions. Though the symposium had not been planned with a theme, the contributions of the speakers complemented each other very nicely. For the proceedings, we have asked the speakers to make their contributions as pedagogically readable as possible and we think that they have largely succeeded. The result reviews a wide but related set of current interests in condensed matter theory.

We first present papers based on the talks by L. W. Burch, John Smith and Jack Dow on surface phases, adhesion and Schottky barriers in which the underlying model of the surface is planar. In the next contributions, on the Laplacian roughening model, equilibrium crystal shapes and dendritic growth by David Bruce, Michael Wortis and Eschel Ben-Jacob respectively one studies decidedly non planar interfaces. This leads naturally to descriptions of material in the talks by Joseph Straley, Rajiv Kalia, and P. L. Taylor on structure of defects in glasses, two dimensional glasses and polymers. Finally, we have papers on the classical dynamics (in networks by Michael Thorpe and in diffusion on fractals by Hisao Nakanishi) and the electron dynamics (based on talks by C. M. Soukoulis, Barry Kunz, Mike Lee, David Stroud, and Craig Davis) of disordered systems.

The symposium is an informal conference designed for the pleasure and enlightenment of the participants and particularly for the students and young investigators among them. It is our hope that these proceedings will serve similar purposes for their readers.

We are grateful to the Argonne Universities Association, the Graduate School and Physics Department of the University of Minnesota and Argonne National Laboratories for supporting the symposium financially. The symposium also owed much to Zlatko Tesanovic and David Price, who served on the local committee, to the participants including especially the more than 25 who contributed to the poster session and to Lockwood Carlson of 3M company, who discussed the role of theoretical physics in industry in a well received afterdinner talk.

J. WOODS HALLEY
March 12, 1985

Minneapolis, Minnesota

STRUCTURE OF THIN PHYSISORBED LAYERS

L. W. Bruch
Department of Physics; University of Wisconsin-Madison
Madison, WI 53706

ABSTRACT

This paper is based on a talk presented at the 12th Midwest Solid State Theory Symposium. It gives an overview of results on the structure of thin layers of inert gases and small linear molecules presented in several recent papers by the author and his co-workers. Uniaxial registry monolayers and the theory of the monolayer to bilayer solid transition are treated. Some qualitative remarks to place the work in the context of other work on multilayer growth and on registry effects in monolayers are included. The work of other authors is not systematically reviewed but references to several of the major papers are included.

1. INTRODUCTION

Thin physisorbed layers of inert gases and small linear molecules--one, two, and three-layer films--show a wide range of phenomena and may be well-suited for a systematic study of modes of crystal growth [1]. There are many examples where the adsorbate- adsorbate interactions in the form known from study of the bulk phases contribute the major part of the lateral cohesive energy of the adsorbed film [2]. There are also examples [Xe and Kr on Pd(100)] where the dispersion force attractions seem to be thoroughly disrupted or compensated in the first layer [3]. The competing periodicities of the "intrinsic" adsorbed lattice, which would be formed on an atomically smooth substrate, and of the actual substrate surface lead to a variety of registry and near-registry structures [1,4].

We will follow the evolution of several adsorbate/substrate combinations from condensation of the first layer to the formation of the 2-layer and 3-layer films. The simplest pattern is observed for Xe on the (111) face of the face-centered cubic metal silver, Xe/Ag(111) [5,6]. The monolayer lattice

constant varies by a small but measurable amount before the bilayer is formed
and the lateral stresses arising from the compression make a large contribution
to the monolayer chemical potential [7]. Indeed, when the compressibility is
explicitly included in the analysis, we arrive at a qualitative understanding
of the similarity of the bilayer condensation chemical potential for a given
adsorbate on several substrates [3,5,8]. Although one might have viewed the
first layer holding potential as setting the "origin" or "zero" of the chemical
potential, the chemical potential at the bilayer formation is only weakly
dependent on the first layer holding potential.

There is much interest in the approach to bulk properties as thicker films
are formed [9-17]. For the system Xe/Ag(111) [5] the lateral nearest neighbor
spacing L of the Xe atoms at monolayer condensation was 2% larger than the
spacing L_s in bulk Xe at its sublimation curve at the same temperature. At and
beyond the second layer formation the spacing L was indistinguishable from L_s,
i.e., they were equal to 0.2%. This led to a belief [5,7] that for Xe/Ag(111)
one could grow arbitarily thick films continuously, that Xe/Ag(111) followed
the Frank-van der Merwe mode of crystal growth [14]. In some other adsorption
systems involving inert gases and small molecules the continuous growth is
apparently not achieved [15,16]. The layer-plus-island, or Stranski-Krastanov,
growth mode [14] is proposed to occur for several linear molecules on the basal
plane surface of graphite (Gr) and for adsorbed layers with large quantum
effects (Ne/Gr and He/Gr) [15]. A qualitative discussion can be given [14] of
effects which enhance the likelihood for the Stranski-Krastanov mode to occur:
"almost any factor which disturbs the monotonic decrease in binding energy,
characteristic of layer growth, may be the cause." Structural considerations
provide a guide: registry, symmetry, and compression each play a role. The
model calculations of monolayer structures will be discussed here in this
context.

A related subject, which we do not treat here, is the development of
electronic structure in thin metal overlayers [17].

The presentation is organized in four parts. In Section 2 the phenomena
for two-dimensional solids, i.e., monolayer solids adsorbed on smooth planar
surfaces, are summarized. Experimental systems which exhibit such intrinsic
monolayers include Xe/Ag(111) [5] and Ar/Gr [18]. In Section 3 the monolayer
packing problems for molecules with shapes, such as linear molecules
[19,20,21,22], are discussed. In Section 4 registry and near-registry
monolayers on adsorbing surfaces with rectangular symmetries are discussed.
This includes systems such as Xe/Ag(110) [23], Kr/Cu(110) [24,25], and
Xe/Cu(110) [23,25] for which the observations of registry structures have been
used [26] to estimate the lateral corrugation of the holding potential.
Finally in Section 5 recent work and speculations [27] for compressed
monolayers are discussed, including the special features arising for adsorbed
helium as a consquence of the large variations in the holding potential between
substrates [28,29,30].

2. Two-Dimensional inert Gas Solids

In this section we discuss the observations on and modelling of the
intrinsic, close-packed monolayer solids of physisorbed inert gases. As is the
case for the three-dimensional solids [31], the inert gases condense in rather
simple structures in which most of the cohesive energy is the result of
pairwise interactions. Knowledge of the pair potential from the dilute

three-dimensional gas phase transfers to the 3D solid and, in large part, to the adsorbed layers [2]. Also in analogy to the case of the three-dimensional solids, there is a strong interplay in the analysis between the development of the statistical mechanical theory for calculating thermal properties and the determination of the full interatomic potential as modified by substrate-mediated processes in the adsorption. The experimental systems which are studied by diffraction on single crystal substrates and which most closely follow the model of a monolayer solid on a smooth dielectric continuum adsorbing surface are Xenon, Krypton, and Argon adsorbed on Ag(111) [5,6,32] and Argon and Neon adsorbed on Gr [18,33,34].

A characteristic of the physisorbed monolayers is that a continuous variation of the adlayer lattice is experimentally possible [5]. The layer can be manipulated by adjusting the temperature and pressure of coexisting 3D gas and its response (e.g., compressibility) can be measured. At monolayer condensation, which occurs at the 2D sublimation curve, the diffraction experiments provide values for $L_o(T)$, the lattice constant of the essentially unconstrained (zero spreading pressure) monolayer solid, and also measures of the density of the coexisting 2D gas at temperatures not far from the supposed 2D triple point. A thermodynamic path followed in such experiments is shown in Figure 1, which is the phase diagram of Xe/Ag(111) [5]. The path ABC, at constant pressure and decreasing temperature, goes from condensation from a dilute 2D gas to a 2D solid (at line a) through a region of monolayer solid (along AB) to the condensation of bilayer solid (at line b) and then on to the appearance of 3D solid (at line c, the bulk sublimation curve). Support for this interpretation of the state of the adsorbed layer is shown in Figure 2, taken from the work of Unguris [5]: Figure 2a shows the adsorbate coverage/area derived from the attenuation of a diffraction signal from the underlying substrate, while Figure 2b shows the combination of these data with the measured lateral nearest-neighbor spacings L in the form of an adsorbed area (coverage times L^2). The slight slope to the long plateau in Figure 2a (region AB of Figure 1) is absent in Figure 2b and the adsorbed area at B, bilayer formation, is twice that along the plateau, to 1%. This analysis led to the identification [5,7] of the Xe/Ag(111) few-layer solids as particularly clean experimental systems with very few vacancies. The slight rise before the monolayer condensation in Figure 2a corresponds to a 2D gas of density 10% of the monolayer solid, an intriguing gas which has been discussed at length elsewhere [35].

The bilayer formation can be viewed as the condensation, under increase of chemical potential, of the dilute thermally activated second layer gas which is present even for the monolayer solid. In computer simulations [36] of the Xe/Gr system at temperatures above 100 K, such gas of density more than 10% of the first layer solid is seen. Modelling of the lower temperature solids represented by Figure 2 has the simplifying feature that the second layer gas density prior to bilayer formation is quite dilute. The statistical mechanical theory of the monolayer to bilayer transition at intermediate temperatures includes effects of vacancies in the solid layers and of the second layer gas of the monolayer solid [37].

A consequence of the compressibility of the monolayer solid is that there is not a unique monolayer lattice constant: for Xe/Ag(111) the lattice constant at monolayer condensation is 1.8% larger than the near-neighbor spacing L_s in the 3D solid at the bulk sublimation curve, while at the bilayer formation the lattice constant is within 0.2% of L_s [5]. With the precision being achieved in measurements of the lattice constant [5,6], it is important to compare

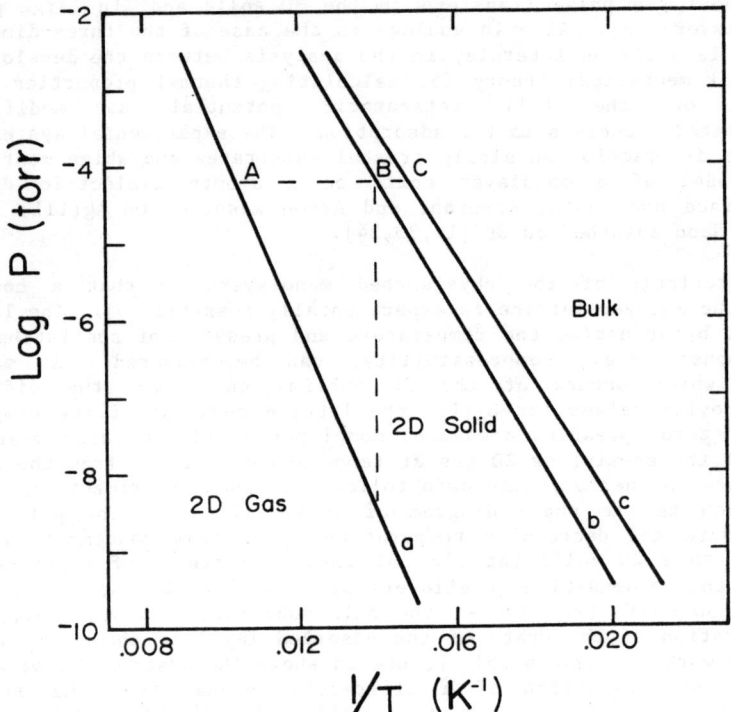

Figure 1. A portion of the phase diagram for Xe/Ag(111). The coordinates are the pressure p of a coexisting 3D gas and temperature T. The dashed lines denote paths of isobaric compression (ABC) and isothermal compression.

results at the same state of the phase diagram. In particular, small differences in the coverage (a few percent) can make large difference in the compression.

The statistical mechanical theory of the monolayer solid has been developed in various approximations [38-42]. The adatom-adatom interactions include the pair potential known from the analysis of 3D phases and several substrate mediated interactions. The parameters for the interactions of adsorption induced dipoles and for the substrate modifications of the London-van der Waals dispersion forces, the Sinanoğlu-Pitzer-McLachlan interaction, are fairly well determined for the adsorption on Ag(111); this has been reviewed elsewhere [2]. The success of the statistical mechanical modelling of the Xe/Ag(111) adsorption [40] has been taken to mean that no large terms have been omitted from the interactions there. For Xe/Pd(100) the observed phenomena are grossly different [3] from the Xe/Ag(111) case; estimates of the contributions of the adsorption dipoles and the McLachlan

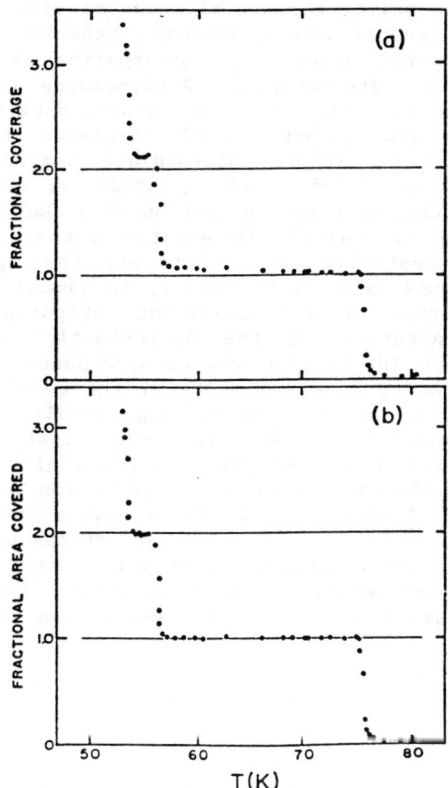

Figure 2. a) Fractional Xe coverage on Ag(111) for a flux of 3D gas at effective pressure p = 3×10^{-8} torr. b) Fractional area, formed by combining the data of (a) with the measured lattice constant of the Xe adlayer, Ref. 5.

interactions do not lead to large enough effects to account for the difference. While one approach has been to argue for the importance of finite geometry effects [43] in this case, it is also plausible that additional substrate-induced processes leading to substantial modification of the overlap interaction of Xe atoms arise because of the larger chemical activity of the palladium substrate. Even if the interaction model is known, the observations of the monolayer solids cover temperature ranges with substantial thermal excitation and the evaluation of the statistical mechanics itself becomes complex.

A relatively simple limit of the statistical mechanics is at low temperatures where a small amplitude oscillation approximation [39] for the atomic motions is accurate except for the light inert gases. This is the quasiharmonic approximation, using the harmonic lattice dynamics; the special-points evaluation of Brillouin zone sums [44] makes it a very rapid calculation. The quasiharmonic approximation has the advantage of including collective motions of the solid, the normal modes of vibration, and leads to a Debye theory at very low temperatures. However, experience with extending the theory to include anharmonic effects by perturbation theory at intermediate temperatures has been rather discouraging. A surprising but useful result from several studies [45,46] has been that an approximation which treats the anharmonicity in some detail, but largely neglects the correlations is frequently quantitatively useful at intermediate temperatures. This is a version of the Einstein model of a solid, based on the Lennard-Jones and Devonshire cell theory [45], in which the motion of a central atom in the field of its fixed neighbors is calculated exactly without recourse to small amplitude displacement approximations. Although the approximation can be formulated as a variational mean field theory, in practice it has frequently been applied [40,46] without a self-consistent determination of the average cell potential. The accuracy of the approximation was established in calculations for 2D Xenon interacting via Lennard-Jones (12,6) potentials at low temperatures by comparing to the results of the quasiharmonic theory [39] and at high temperatures, approaching melting conditions, comparing to the results of Monte Carlo simulations [46,47]. An application of the cell model for cases such as neon was developed [42]: in place of a classical continuum evaluation of the cell partition function, the partition function was evaluated from the quantum mechanical energy levels for motion in the cell potential. A technical difficulty in developing the numerical methods was eventually traced [42] to the fact that the error-determining step in a matrix formulation of the Schrödinger equation in mathematical 2D is different than in 3D; it is the approximation to a boundary condition at zero separation in 2D.

Table 1

Monolayer Solids

	Ne	Ar	Xe
$\varepsilon(K)$ [a]	37	120	230
$r_o(Å)$ [a]	3.13	3.82	4.50
Λ^*	0.58	0.19	0.06
L_o/r_o	1.050	1.007	0.9956
E_o/ε	-2.25	-3.01	-3.25
$K_T(\varepsilon/r_o^2)$	0.0303	0.0177	0.0144

[a] Energy and length scales of Lennard-Jones (12,6) potentials.

A summary of properties calculated for Lennard-Jones models of monolayer solids is presented in Table 1. The pair potential for atoms at separation r is

$$\phi(r) = \varepsilon\{(r_o/r)^{12} - 2(r_o/r)^6\} \qquad (2.1)$$

and the de Boer parameter is defined by

$$\Lambda^* = (h/r_o \sqrt{m\varepsilon})2^{1/6}, \qquad (2.2)$$

where h is Planck's constant and m is the atomic mass. The reduced monolayer lattice constant at zero spreading pressure, L_o/r_o, the ground state energy E_o/ε and the dimensionless compressibility are listed there for conventional parameters for Ne, Ar, and Xe. For comparison, the corresponding values for pure classical solids ($\Lambda^*=0$) are $L_o/r_o= 0.97$ in 3D and $L_o/r_o=0.9902$ and $E_o/\varepsilon=-3.382$ in 2D. The 2% expansion in L_o from 3D to 2D, which is purely a consequence of lattice geometry in this calculation, is in fortuitous agreement with the expansion observed for Xe/Ag(111). That expansion is now known [48] to arise from the combination of several partially offsetting 1% effects to the determination of L_o.

The entries for Neon [49] in Table 1 show the effects of large zero-point energy effects. Attempts to develop a perturbation theory of the monolayer solid show the dilation is so large that there are substantial anharmonic effects [50]. The large compressibility, also an effect of the dilated lattice, is associated with a large compression of the monolayer between its condensation and second layer formation. The change in lattice constant is found [49] with various models for Ne/Gr to be 4 to 5%, or double the change for Xe/Ag(111). The magnitude of the lattice constant at bilayer formation was calculated to be close to the measured near-neighbor spacing of 3D Ne at the sublimation curve L_s. Experiments found it to be 1 to 2% smaller than L_s [33,34]; part of the difference is related to uncertainty in the length scale of the 3D pair potential [51]. This discrepancy has some consequence for the understanding of the crystal growth mode of Ne/Gr. The Ne/Gr system has been proposed to follow a Stranski-Krastanov growth mode, with the driving factor being the mismatch in lattice constant between the compressed monolayer and the 3D solid at sublimation [10,16]. According to this argument it is likely that the compressed monolayer solids of D_2/Gr and H_2/Gr also follow this growth mode because they show even larger compressions relative to their 3D solids [52].

There are, however, competing effects which may change the behavior. The monolayer of Xe/Gr at its limit of compression [8] has, at 60 K, a value of L which is 2.5% smaller than L_s; however the bilayer then "springs back" and has an L value close to L_s, a phenomenon of "inverted epitaxy" [53]. This could occur for other strongly compressed layers. Second, the over-compression of the quantum layers is related to their relatively large compressibility; the large compressibility would also lower the energy needed to form dislocation-like lattice defects to heal the lattice toward the bulk structure [9]. These are structural processes which may be difficult to infer without detailed diffraction studies of the few-layer system.

3. Linear Molecules

For the intrinsic inert gas monolayer the atomic packing problem is simple and the structure is the close-packed triangular lattice; there is no other 2D Bravais lattice close in energy to this except for a degenerate limit of the centered rectangular lattice. Even for the double layer, which is a close-packed stack of two triangular lattices, the structure to be studied is readily apparent. At the trilayer, the question of the stacking sequence arises [7]: whether it is the beginning of the hexagonal close-packed or face-centered cubic sequence of the 3D solids. The energies of the two 3D structures are very close in magnitude and it is conceivable that the external potential provided by the substrate might lead to different initial stacking sequences than the bulk. Some phonon spectra for the trilayers of Ar,Kr, and Xe on Ag(111) are now available from inelastic helium atom scattering [6], but the resolution is not high enough to use the observed dispersion relations to distinguish between the stacking sequences.

The packing problem for thin layers of small linear molecules is much more complex. In the Pa3 crystal structure of 3D solids of N_2 and of CO_2 [54] the molecules are not flat in the close-packed plane of the solid and the solid consists of 4 sublattices. Under compression the N_2/Gr monolayer may form a pinwheel structure [20] with molecules oriented at angles to the adsorbing plane. However, in most calculations to determine the configuration at monolayer solid condensation, two interpenetrating rectangular lattices of different molecular orientations, a herringbone pattern [21,22], are considered as the most likely structure. The determination of the minimum energy lattice then includes electrostatic interactions of permanent quadrupole moments of the molecules (and the effect of substrate screening charges) and the overlap repulsions and van der Waals attractions among the admolecules [22]. The overlap repulsions lead to angles for the herringbone array which differ from the angle which minimizes the energy of the quadrupoles on opposite sublattices.

For two quadrupoles of moment Q and separation R oriented so that their axes are coplanar with the separation vector the configuration of lowest electrostatic energy Q has the axes perpendicular to each other

$$\phi_Q = -3Q^2/R^5. \quad (3.1)$$

There is another coplanar configuration with electrostatic energy close to this

$$\phi_Q = -(18/7)Q^2/R^5, \quad (3.2)$$

where the angle θ is the inverse cosine of $\sqrt{3/7}$. The highest energy configuration has

$$\phi_Q = 6Q^2/R^5. \quad (3.3)$$

The fact that Eq.(3.1) and (3.2) are so close in magnitude means that both sets of near neighbors in a planar herringbone lattice can have nearly minimal electrostatic interaction energy [55].

The quadrupolar energies are a major portion of the cohesive energy of 3D solid CO_2 [56]. Thus, the experimental proposal [19] that CO_2 does not wet graphite at low temperatures, even to the extent of forming a monolayer solid, stimulated a model calculation [22] of the ground state energy of the monolayer solid including the electrostatic screening response of the substrate. As in the evaluation of the substrate-mediated dispersion energies [2], ideal screening response to external charges was used at very small separations, where it clearly becomes a rough approximation to the substrate response. The result of the calculations, for a two-sublattice model with the CO_2 lying in the adsorbing plane, was that the ground state solid was likely to be incommensurate with the graphite lattice; this differed from a proposed registry structure inferred from the nominal coverage in the experiment. The calculation showed that the substrate screening of the molecular quadrupole moments had a large effect in the ground state energy and that it was marginal whether the model gave wetting. Both these results showed that to make progress in modelling the CO_2/Gr system more knowledge of the holding potential and of its dependence on the angle of the CO_2 axis to the surface normal were needed.

The focus of the model calculation [22] for CO_2/Gr was on the screening response of the substrate and its contribution to the electrostatic portion of the adlayer energy. However there is a second perspective [14] which also suggests the difficulty of a smooth growth of the adsorbed film: the bulk solid in this orientation has close-packed planes where the molecules are at angles relative to the plane [54]. It is difficult to extend smoothly from the likely first layer packing to this type of geometry.

The discussions of bilayer formation in Sec.2 and of the CO_2 layer here both are based on pictures of thin films with well-defined layer structure. In the modelling of the bilayer films at intermediate temperatures a failure of the approximations came [37] when the perpendicular motions of second layer atoms became large and led to poor definition of the second layer. The fact that the CO_2 monolayer film wets Gr at temperatures above 104 K but not at lower temperatures [19] appears difficult to understand in the context of the layered solid picture. The monolayer solid becomes favorable in chemical potential relative to the bulk solid at a high enough temperature, which means that entropy plays a major role. For the monolayer solid to have higher entropy than the 3D solid probably requires significant motions of the adsorbate out of the monolayer plane.

4. Registry Adlayers

The modulation phenomena in monolayer solids adsorbed on strongly corrugated substrate surfaces reflect the competing periodicities of the intrinsic adlayer and of the substrate. In this context one makes contact with general concepts of frustrated interactions and with specific models of the nonlinear response of idealized models such as the Frenkel-Kontorova, Frank-van der Merwe, one-dimensional chain in a periodic potential [57,58,59]. The most extensively studied monolayers showing these phenomena are triangular adlayer lattices on the basal plane surface of graphite [60]. The theoretical analysis has been difficult and complicated because some of the significant modulation patterns are two-dimensional intersecting arrays of misfit dislocations or domain walls [61,62]. However, there are systems where the substrate potential has approximately rectangular symmetry and which appear to be simpler to treat [26]: inert gases on the (110) face of face-centered cubic metals. There is not yet a great deal of data on these systems [23,24,25,63], but the available information has been used [26] to make some estimates of the corrugation of the one-atom holding potential and to identify features in the data which are difficult to understand or surprising in the context of simple models of the adlayer response.

The corrugation of the holding potential for a planar substrate surface is expressed in terms of the Fourier decomposition of the holding potential [64]:

$$V(r,z) = V_o(z) + \Sigma_{g(\neq 0)} V_g(z) e^{ig \cdot r}; \qquad (4.1)$$

the Cartesian coordinate axes have the z-component perpendicular to the surface and the vector r (with x and y components) parallel to the surface plane. The vectors g are the reciprocal lattice vectors of the substrate surface plane. For model holding potentials formed by summing the interactions of adatom-substrate atom pairs [64], the Fourier amplitudes V_g decrease in magnitude rapidly with increasing magnitude of g for weakly corrugated surfaces. Measurements of the scattering of helium atoms from Pd(110) have shown the first few amplitudes decrease rapidly [65].

When the surface lattice of the substrate has a different symmetry, such as rectangular or centered rectangular, than the triangular lattice of an intrinsic monolayer of an inert gas, the description of registry must be generalized from the case of the triangular adlayer on triangular surface [4]. Also, a rectangular surface lattice may drive distortions of the adlayer lattice to a centered rectangular lattice or even to an oblique lattice. Simple registry in these cases arises when elements of the reciprocal lattice of the adlayer coincide with some elements of the reciprocal lattice g of the substrate surface [4]. Coincidence with leading surface reciprocal lattice vectors may also occur after small adlayer distortions to a lattice with more than one atom per unit cell [66], as perhaps is the case for Ne/Gr at very low temperatures [34]. However the net lowering of the adlayer energy for optimized positions in the larger unit cell is expected to be smaller [66,67] than energies arising directly from the leading terms in Eq. (4.1). Detailed analysis has so far only been performed for the case of the adlayer with one atom per unit cell. Also, if the uniaxial registry persists through the monolayer regime, the formation of the bilayer may occur at a chemical potential very close to the bulk value and the structural misfit may make it difficult to achieve an extended layer-by-layer growth. This possibility was examined in a model calculation [27] for uniaxially registered Xe/Cu(110): the calculated bilayer formation was very close to the bulk condensation.

We now discuss model calculations [26] for planar centered rectangular adlayers on rectangular adsorbing lattices. As coordinates of the fcc(110) surface take the x-axis to be along the [001] direction and the y-axis along the [1$\bar{1}$0] substrate is ℓ, the distance between nearest neighbors at the (110) surface is $\ell\sqrt{2}$ in the [001] direction and ℓ in the [1$\bar{1}$0] direction. There are close-packed troughs on the surface running parallel to the y-axis. Uniaxially registered lattices where the atoms of a centered rectangular adlayer reside in these troughs are observed for Xe/Ag(110) [23], Xe/Cu(110) [23,24], and Kr/Cu(110) [24,25]. The lateral interactions of inert gas atoms such as Xe and Kr in island-forming systems are fairly well known [2,8,38]. Thus the occurrence of the registry lattices can be used to make estimates of the amplitudes of the corrugation potential in a truncated series corresponding to Eq. (4.1):

$$V(r,z_o) = V_o + 2V_1 \cos(2\pi x/\ell\sqrt{2}) + 2V_2 \cos(2\pi y/\ell). \qquad (4.2)$$

The amplitudes V_1 and V_2 must be large enough that it is energetically favorable to distort the adlayer lattice from the intrinsic uniform triangular lattice to the (uniaxially) registered structures. A first estimate [26], which provides a lower bound on the critical potential amplitude, is to compare the value of the lateral potential energy in the two structures. Indeed, although this comparison does not explicitly treat the energy of modulated adlayer lattices of small misfit relative to the registry lattice, it agrees with the result of the more detailed calculation to about 20%.

The detailed calculation [26] is a version of the Frank-van der Merwe calculation for a harmonic linear chain in an external sinusoidal potential [57,60]. The average centered rectangular lattice is modulated with harmonics of the misfit wave vector (the smallest reciprocal lattice vector difference between adlayer and surface lattice). The dispersion relation of the adlayer normal modes, for a Lennard-Jones (12,6) pair potential between adatoms, is included and the interaction energy with the periodic substrate potential is evaluated as a continuum average, which omits the pinning potential [66] for dislocations on a discrete surface. The amplitudes of the positional modulations are adjusted to minimize the total energy [59]. Anharmonic effects enter in two partially offsetting ways. First, the harmonic elastic constants for inert gas adlayers are very rapid functions of the lattice constant, as noted by Venables [61]. Second, for the rather dilated lattices which occur for the C(2X2) lattice of Xe/Cu(110) and for Xe/Ag(110) and for the compressed (along [001]) spacings of Xe/Cu(110), the nonlinear response of the adlayer is important in limiting the size of the modulation amplitudes. The sort of pathology which can occur if the anharmonic response is not included is shown in model calculations [26] to determine the value of V_2 needed to stabilize the C(2X2) lattice of Xe/Cu(110) relative to the optimized uniaxial registry lattice (registry only along the x-axis). The dilation in the [1$\overline{1}$0] direction relative to the optimized uniaxial lattice is 6% to 10%, depending on the lateral interaction model; it is so large that the harmonic lattice constants become so small for displacements along the [1$\overline{1}$0] direction that states of finite misfit occur where essentially all the adatoms are in registry positions and the misfit is taken up in a small fraction of an interatomic spacing. That is, nearly the full registry energy is available at small finite misfit because the domain walls are very narrow; the cubic and quartic anharmonic perturbation theory terms both restore the situation of broader domain walls.

Another method to estimate the critical values of V_1 and V_2 to stabilize the registry structures is [26] to use the results of the continuum Frank-van der Merwe theory [57] for a linear harmonic chain, with the adlayer elastic constant as a parameter. As just noted, when the registry lattice is quite dilated relative to the intrinsic triangular lattice the harmonic elastic constants are small; inclusion of anharmonic effects increases the effective elastic constants. Also, if the registry structure is strongly compressed the stability of the registry structure requires a more detailed comparison of lattices at distinctly different lattice constants, since the onset of the registry as a function of V_1 can be [26] a "first order transition." In examples of both types the energy balance argument was found to be a more accurate guide to the requirements for registry than the Frank-van der Merwe theory. This happened in spite of the fact that there are sizable fractions of the registry energy included in the energy of the layer at 1% misfit in these calculations. Thus the "all-or-nothing" estimates of registry energy used for the discussion of stability in other work [68] may be less crude than had been thought.

The estimates of the corrugation amplitudes of the holding potential, V_1 and V_2, on the fcc(110) surface, can be compared with the estimates available for Xe and Kr on Gr [69,70]. The lower bounds for V_1 of Xe/Cu(110) and Xe/Ag(110) which were obtained [26] are noticeably larger than the estimates [70] of the leading amplitude for Xe/Gr; in view of the topography that was a plausible result. The value of V_2 for Xe/Cu(110), where experiments [24] show that the Xe can be compressed from C(2X2) registry with a small increase in chemical potential, was estimated to be twice the value of the leading amplitude for Xe/Gr; the sense of the difference is consistent with the magnitudes of the corresponding reciprocal lattice vectors. Surprisingly, the estimate for V_1 for Kr/Cu(110) is smaller than the leading amplitude for Kr/Gr [69]. The small value was set by a report [25] that the Kr/Cu(110) adlayer could be compressed out of uniaxial registry with a small change in the average lattice constant. More detailed experimental studies are certainly desirable.

The modulations of the adlayer at finite misfit relative to registry structures are, in principle, observable in diffraction experiments. The only system for which an observation of the modulation sideband has been reported is the Kr/Gr system near the $\sqrt{3}$ R 30° commensurate phase [71]. It has been a puzzle that only one of the modulation sidebands for this system has been observed, but there is a model calculation [69] for a static lattice which apparently generates a nearly one-sided modulation, in qualitative agreement with the experiments [71].

A centered rectangular uniaxially registered adlayer lattice has been observed on a triangular substrate surface for the system CF_4/Gr [72]. This "striped" structure was identified in X-ray diffraction from a "powder" by a characteristic 2:1 intensity ratio of slightly split diffraction rings. There is also a triangular lattice of CF_4/Gr which shows the Novaco-McTague rotation [74]. The observations of the rotation include small misfits relative to the 2 X 2 registry lattice, for which a modification of the Shiba calculation [75] might be appropriate. However, a likely adsorption site for the CF_4 on Gr is an "atop" site, above a surface carbon atom [76], rather than the honeycomb center site for Kr/Gr. The sign of the leading Fourier amplitude in Eq.(4.1) is then opposite to its sign for the Kr/Gr case treated by Shiba [75] and it is unclear how to scale his results to the CF_4/Gr case. In contrast to Eq.(4.2), the overall sign of the leading Fourier amplitude for the triangular lattice is not reversed by a simple translation of the origin of the x and y coordinates. Also, there is the question of whether there is a dislocation-rich 2D fluid [77] intermediate between the commensurate structure and the incommensurate aligned and rotated solid and what effects similar dislocations would have in the 2D solids.

There are only a few cases where registry or partial registry structures have been observed near the monolayer-to-bilayer transition. For Xe/Gr [8] there are temperatures where the transition occurs from a $\sqrt{3}$ R 30° commensurate monolayer to a triangular bilayer stack with lattice constant close to the bulk value. The uniaxially registered Xe/Cu(110) lattice was compressed with a chemical potential increase large enough [24] to nearly achieve bulk condensation. These are examples where there might be large structural effects on the possible layer-by-layer growth. The Xe/Gr system, though, has been identified [15] as following the extended layer growth mode. There are a few other brief reports of adsorbed phases beyond a monolayer registry lattice [63] but investigation of those phases with crystal growth mechanisms in mind has not yet been reported.

5. Compressed Monolayers

In closing we return to the topic of compressed monolayers and the relation of the compression to the value of the one-atom holding potential [27], discussing two points: (1) the relation of the first-layer holding potential to the chemical potential at second layer condensation; and (2) the special features which may arise in comparing the adsorption of helium on several substrates.

For simplicity and explicitness, consider the case of the monolayer-to-bilayer transition of a classical adlayer at zero temperature. Then the energies that enter in the analysis are potential energies: (1) the first layer one-atom holding potential ε_o, (2) the interaction energy with the substrate of an atom in the second layer of the bilayer ε_{II}, (3) the lateral adatom energy in the monolayer $\varepsilon_{\ell 1}$, and (4) the adatom-adatom interaction energy in the bilayer $\varepsilon_{\ell 2}$. In terms of potential energies Φ_{ij} for interactions among adatoms in layers i and j the energies $\varepsilon_{\ell 1}$ and $\varepsilon_{\ell 2}$ are

$$\varepsilon_{\ell 1} = \Phi_{11}$$
$$\varepsilon_{\ell 2} = \Phi_{11} + \Phi_{22} + \Phi_{12} \quad . \tag{5.1}$$

The internal energy per adatom of the monolayer is

$$u_1 = \varepsilon_o + \varepsilon_{\ell 1} \tag{5.2}$$

and for the bilayer it is

$$u_2 = \frac{1}{2}(\varepsilon_o + \varepsilon_{II} + \varepsilon_{\ell 2}) \quad . \tag{5.3}$$

If the areas per adatom in the monolayer and bilayer stacks are a_1 and a_2, respectively, the zero temperature spreading pressures and chemical potentials (enthalpies) are

$$\phi_i = -du_i/da_i,$$

$$\mu_i = u_i + a_i\phi_i. \tag{5.4}$$

At the monolayer-to-bilayer transition the conditions of mechanical and mass transfer equilibrium are

$$\phi_1 = \phi_2$$

$$\mu_1 = \mu_2 \tag{5.5}$$

so that the chemical potential there is

$$\mu_{II} = [1 - \frac{1}{2}\frac{a_1}{a_1-a_2}]u_1 + \frac{1}{2}\frac{a_1}{a_1-a_2}(\varepsilon_{II}+\varepsilon_{\ell 2}-\varepsilon_{\ell 1}). \tag{5.6}$$

Now, for Ar, Kr, and Xe on Ag(111) [5,6,32] and for Ar and Kr on Gr [18,78], there is no discontinuity observed in the lateral nearest-neighbor spacing L at the monolayer-to-bilayer transition. Writing this result as

$$a_1/a_2 \simeq 2, \tag{5.7}$$

the value for μ_{II} is

$$\mu_{II} \simeq \varepsilon_{II} + (\varepsilon_{\ell 2}-\varepsilon_{\ell 1}). \tag{5.8}$$

For Xe/Gr a discontinuity of 2 1/2% at the transition was observed [8] at around 60 K; then the coefficient of the first term on the right-hand side of Eq.(5.6) is -.05 rather than zero and there is a positive contribution to μ_{II}.

To the extent that the discontinuity in the spacing L is small at the bilayer formation the potential energies Φ_{ij} in Eqs.(5.1) are evaluated at the same lateral spacing and the value for μ_{II} is

$$\mu_{II} \simeq \varepsilon_{II} + \Phi_{22} + \Phi_{12}. \tag{5.9}$$

There is no direct dependence of μ_{II} on the first layer holding potential or on first layer adsorption induced interactions in this approximation. There is an indirect dependence on these terms because they are part of the stress balance determining the change in lattice constant L from the value at monolayer condensation. Even with the small discontinuity reported for Xe/Gr, the fractional direct contribution of ε_o to μ_{II} is small. It acts to bring the value of μ_{II} closer to the value μ_o for bulk condensation than the value for a system such as Xe/Ag(111); the difference in the polarization potential energy μ_{II} for Xe on Gr and on Ag [2,79] partially compensates this term, and the values of μ_{II} for Xe/Ag(111) and Xe/gr are similar [5,80]. The insensitivity of μ_{II} to the first layer adsorption-induced interactions is shown in data for Xe/Pd(100) where the bilayer forms [3] at a chemical potential close to μ_{II} for Xe/Ag(111).

The result of this discussion is close to the conditions assumed in the Singleton-Halsey theory of multilayer adsorption [81]. For classical adsorbates such as Xe the spacing L at the bilayer condensation is close to that of the bulk 3D solid at its sublimation curve at the same temperature; there are only small changes in L as thicker films are formed. An extension of these considerations to the bilayer-to-trilayer transition gives a chemical potential displacement relative to μ_o which depends directly on the third-layer atom polarization potential with the bulk substrate and a difference in the adatom-adatom potential sums relative to the extended sums in the bulk. What is not discussed here is the difference in entropy contributions and, as noted in the discussion of CO_2/Gr in Sec.3, they may lead to thermal variation of the layer stability relative to bulk. However, as might be expected, analysis [32] of a lattice gas model (the de Oliveira-Griffiths model [82]) of multilayer formation also leads to the Singleton-Halsey result at low temperatures where entropy effects in the dense layers are small.

The role of the compressibility in causing a very reduced dependence of μ_{II} on the first layer holding potential ε_o has been shown in special cases by model calculations. Most recently [27], in a study of the bilayer formation for uniaxially registered Xe/Cu(110), a change of the first layer holding potential by 400 K led (for the same lateral interactions and ε_{II}) to changes of less than 2 K in μ_{II}. However, the increment in chemical potential from monolayer condensation to bilayer formation varies directly with ε_o and this was used as part of the anlysis [27] which led to a proposed revision of the heat of monolayer condensation of Xe/Cu(110). The stability of higher layers of a film relative to bulk clustering depends on chemical potential differences which are a small fraction of the total chemical potential; constraints on the structure imposed by substrate registry effects can readily contribute on this scale. Organized large-scale relaxations in the multilayer film by lattice defects such as stacking faults can heal the structure but they are effects which are complicated to include in some of the more algebraic formulations of the multilayer theory.

The range of the monolayer (2D) phase diagram which is accessible for a given adsorbate is related to the difference $\Delta\mu$ between the chemical potentials at first and second layer condensations: the evolution of the monolayer melting curve with increasing temperature depends on $\Delta\mu$. If the process is viewed as solidification under an increase in the spreading pressure, then clearly there can be a competition between the solidification and the second layer formation [83]. The melting line for monolayer Xe/Gr has been followed experimentally for chemical potential increases of up to 2/3 of $\Delta\mu$ [84].

The analysis of the dependence of the chemical potential at second layer formation on the first layer atom-to-substrate binding energy ε_o has been extended to the case of helium adsorption [27]. For ^{4}He/Gr, at second layer condensation the second layer atoms form a fluid atop a solid first layer [85]. This requires a change of the argument based on Eq.(5.7), but a cancellation similar to Eq.(5.8) is again obtained. However, the monolayer quantum solid has a large lateral compressibility compared to the classical solids, so the lateral spacing L varies considerably with ε_o. The fractional variation of ε_o for helium among substrates is particularly large: from 12 meV for ^{4}He/Gr [28] to 6 meV for ^{4}He/Au(110) [29] to 1 meV for ^{3}He/H$_2$-film [30,86]. For the small holding potentials, only a very limited portion of the mathematical 2D phase diagram of a quantum fluid may be accessible [87]: for ^{3}He in three dimensions the chemical potential increase from condensation of liquid to solid formation is 1 mev at 0 K and 2 meV at 2 K. There are probably unusual structures still to be found for adsorbed helium and, perhaps, for adsorbed hydrogen.

6. CONCLUDING REMARKS

Monolayer solids of inert gases on metals occur in a variety of structures, but at the onset of the bilayer solid, the difference from the 3D structure has usually become small. The persistence of structures which are stabilized by registry energies with the substrate leads to exceptions to this generalization, but the compressibility of the adlayer ordinarily permits the structure to heal to an extent that there are only small mismatches between the thin film of inert gas and the bulk structure. The chemical potential differences between the possible phases are also small then, so that accurately identifying and evaluating the competing processes which determine the mode of crystal growth may be difficult. Precise structural investigations are likely to be an important part of the experimental determination of the evolution of the thin layers. The theoretical considerations on thin solid layers described here are of most use at low temperatures where the layer structure is well-defined and vacancy effects in the solids are small. At present, there are a few systems where the experimental data are for conditions where such simple modelling is accurate and can be carried through to give explicit accounts of some of the competing processes. Several lines of research now converge on the thin inert gas layers and the multiplicity of approaches may be quite productive.

REFERENCES

1) Interfacial Aspects of Phase Transformations, B. Mutaftschiev, ed., (Reidel, Dordrecht, 1982); Phase Transitions in Surface Films, J. G. Dash and J. Ruvalds, eds., (Plenum, New York, 1980); Statistical Mechanics of Adsorption, M. W. Cole, F. Toigo, and E. Tosatti, eds.,(Surface Science Volume 125, North Holland, Amsterdam, 1983).

This work was supported in part by the National Science Foundation through grant DMR-8214518.

2) Bruch, L.: Surface Sci. (1983) 125, 194 and references contained therein.

3) Moog, E. R. and Webb, M. B.: Surface Sci. (in press).

4) Bruch, L. W. and Venables, J. A.: Surface Sci. (in press).

5) Unguris, J., Bruch, L. W., Moog, E. R., and Webb, M. B.: Surface Sci. (1979) 87, 415.

6) Gibson, K. D. and Sibener, S. J.:(to be published).

7) Bruch, L. W., Unguris, J., and Webb, M. B.: Surface Sci. (1979) 87, 437.

8) Schabes-Retchkiman, P. S. and Venables, J. A.: Surface Sci. (1981) 105, 536.

9) Pandit, R., Schick, M., and Wortis, M.: Phys. Rev. B (1982) 26, 5112; Gittes, F. T. and Schick, M.: Phys. Rev. B (1984) 30, 209.

10) Muirhead, R. J., Dash, J. G., and Krim, J.: Phys. Rev. B (1984) 29, 5074.

11) Goodstein, D. L., Hamilton, J. J., Lysek, M. J., and Vidali, G.: Surface Sci. (in press).

12) Ebner, C.: Phys. Rev. B (1983) 28, 2890; Saam, W. F.: Surface Sci. (1983) 125, 253.

13) Sullivan, D. E.: J. Chem. Phys. (1981) 74, 2604; Cahn, J. W.: J. Chem. Phys. (1977) 66, 3667.

14) Venables, J. A., Spiller, G. D. T., and Hanbücken, M.: Repts. Progr. Phys. (1984) 47, 399 and references contained therein.

15) Seguin, J. L., Suzanne, J., Bienfait, M., Dash, J. G., and Venables, J. A.: Phys. Rev. Lett. (1983) 51, 122; Venables, J. A., Seguin, J. L., Suzanne, J., and Bienfait, M.: Surface Sci. (in press).

16) Bienfait, M., Seguin, J. L., Suzanne, J., Lerner, E., Krim, J., and Dash, J. G.: Phys. Rev. B (1984) 29, 983.

17) Tobin, J. G., Robey, S. W., Klebanoff, L. E., and Shirley, D. A.: Phys. Rev. B (1983) 28, 6169.

18) Shaw, C. G. and Fain, S. C. Jr.: Surface Sci. (1980) 91, L1; (1979) 83, 1.

19) Terlain, A. and Larher, Y.: Surface Sci. (1983) 125, 304.

20) Diehl, R. D. and Fain, S. C. Jr.: Surface Sci. (1983) 125, 116.

21) Pan, R. P., Etters, R. D., Kobashi, K., and Chandrasekharan, V.: J. Chem. Phys. (1982) 77, 1035; see also Fuselier, C. R., Gillis, N. S., and Raich, J. C.: Solid State Commun. (1978) 25, 747.

22) Bruch, L. W.: J. Chem. Phys. (1983) 79, 3148.

23) Chesters, M. A., Hussain, M., and Pritchard, J.: Surface Sci. (1973) 35, 161.

24) Glachant, A., Jaubert, M., Bienfait, M., and Boato, G.: Surface Sci. (1981) 115, 219; Glachant, A., Bienfait, M., and Jaubert, M.: Surface Sci. (in press).

25) Horn, K., Mariani, C., and Cramer, L.: Surface Sci. (1982) 117, 376.

26) Bruch, L. W.: Surface Sci. (in press).

27) Bruch, L. W., Gay, J. M., and Krim, J.: (to be published).

28) Cole, M. W., Frankl, D. R., and Goodstein, D. L.: Rev. Mod. Phys. (1981) 53, 199 and references contained therein.

29) Rieder, K. H., Engel, T., and Garcia, N.: Proc. ECOSS-3, Vol. II p. 861; Cannes, 1980 (Supplement a Le Vide, les Couches Minces, no. 201).

30) Lefevre-Seguin, V.: Thèse d'état, Paris (1984) (unpublished).

31) Klein, M. L., and Venables, J. A., eds.: Rare Gas Solids Vol. I (Academic, New York, 1976) and Vol. II (Academic, New York, 1977).

32) Unguris, J., Bruch, L. W., Moog, E. R., and Webb, M. B.: Surface Sci. (1981) 109, 522.

33) Calisti, S., Suzanne, J., and Venables, J. A.: Surface Sci. (1982) 115, 455.

34) Tiby, C., Wiechert, H., and Lauter, H. J.: Surface Sci. (1982) 119, 21.

35) Unguris, J., Bruch, L. W., and Webb, M. B.: Surface Sci. (1982) 114, 219.

36) Koch, S. W. and Abraham, F. F.: Phys. Rev. B (1983) 27, 2964.

37) Wei, M. S. and Bruch, L. W.: J. Chem. Phys. (1981) 75, 4130.

38) Price, G. L. and Venables, J. A.: Surface Sci. (1976) 59, 509.

39) Phillips, J. M. and Bruch, L. W.: Surface Sci. (1979) 81, 109.

40) Bruch, L. W. and Phillips, J. M.: Surface Sci. (1980) 91, 1.

41) McTague, J. P.: Ann. Rev. Phys. Chem. (1980) 31, 491; Abraham, F. F.: Phys. Reports (1981) 80, 339.

42) Phillips, J. M. and Bruch, L. W.: J. Chem. Phys. (1983) 79, 6282.

43) Soler, J. M., Garcia, N., Miranda, R., Cabrera, N., and Saenz, J. J.: Phys. Rev. Lett. (1984) 53, 822.

44) Cunningham, S. L.: Phys. Rev. B (1974) 10, 4988.

45) Barker, J. A., Henderson, D., and Abraham, F. F.: Physica(Utrecht)A (1981) 106, 226.

46) Phillips, J. M., Bruch, L. W., and Murphy, R. D.: J. Chem. Phys. (1981) 75, 5097.

47) Phillips, J M. and Bruch, L. W.: (unpublished).

48) Bruch, L. W., Cohen, P. I., and Webb, M. B.: Surface Sci. (1976) **59**, 1.

49) Bruch, L. W., Phillips, J. M., and Ni, X.-Z.: Surface Sci. (1984) **136**, 361.

50) Bruch, L. W. and Phillips, J. M.: J. Phys. Chem. (1982) **86**, 1146.

51) Aziz, R. A., Meath, W. J., and Allnatt, A. R.: Chem. Phys. (1983) **78**, 295 and Aziz, R. A.: (private communication).

52) Nielsen, M., McTague, J. P., and Ellenson, W.: J. Physique (Paris) (1977) **38**, C4-10.

53) Webb, M. B.: (private communication).

54) Hirshfeld, F. L. and Mirsky, K.: Acta Crystallogr. A (1979) **35**, 366.

55) Steele, W. A.: J. Physique (Paris) (1977) **38**, C4-61.

56) Murthy, C. S., Singer, K., and Mc Donald, I. R.: Mol. Phys. (1981) **44**, 135 and references contained therein.

57) Frank, F. C. and Van der Merwe, J. H.: Proc. Roy. Soc. (London) A (1949) **198**, 205, 216.

58) Villain, J. in _Ordering in Strongly Fluctuating Condensed Matter Systems_, Riste, T., ed. (Plenum, New York, 1980).

59) McMillan, W. L.: Phys. Rev. B (1976) **14**, 1496.

60) Villain, J. and Gordon, M. B.: Surface Sci. (1983) **125**, 1.

61) Venables, J. A. and Schabes-Retchkiman, P. S.: Surface Sci. (1978) **71**, 27.

62) Abraham, F. F., Rudge, W. E., Auerbach, D. J., and Koch, S. W.: Phys. Rev. Lett. (1984) **52**, 445.

63) Mason, B. F. and Williams, B. R.: Surface Sci. (1983) **130**, 295 and (1984) **139**, 173.

64) Steele, W. A.: Surface Sci. (1973) **36**, 317.

65) Rieder, K. H. and Stocker, W.: J. Phys. C (1983) **16**, L 783.

66) Theodorou, G. and Rice, T. M.: Phys. Rev. B (1978) **13**, 2840.

67) Bruch, L. W.: Surface Sci. (1982) **115**, L 67.

68) Phillips, J. M.: Phys. Rev. B (1984) **29**, 5865.

69) Gooding, R. J., Joos, B., and Bergersen, B.: Phys. Rev. B (1983) **27**, 7669.

70) Joos, B., Bergersen, B., and Klein, M. L.: Phys. Rev. B (1983) **28**, 7219.

71) Stephens, P. W., Birgeneau, R. J., Horn, P. M., Moncton, D. E., and Brown, G. S.: Phys. Rev. B (1984) **29**, 3512.

72) Kjaer, K., Nielsen, M., Bohr, J., Lauter, H. J., and McTague, J. P.: Phys. Rev. B (1982) $\underline{26}$, 5168.

73) McTague, J. P. and Novaco, A. D.: Phys. Rev. B (1979) $\underline{19}$, 5299 and references contained therein.

74) Calisti, S. and Suzanne, J.: (to be published).

75) Shiba, H.: J. Phys. Soc. Japan (1980) $\underline{48}$, 211.

76) Phillips, J. M.: (private communication) and Bak, P. and Bohr, T.: Phys. Rev. B (1983) $\underline{27}$, 591.

77) Coppersmith, S. N., Fisher, D. S., Halperin, B. I., Lee, P. A., and Brinkman, W. F.: Phys. Rev. B (1982) $\underline{25}$, 349 and Pokrovskii, V. L.: J. Physique (Paris) (1981) $\underline{42}$, 761.

78) Fain, S. C. Jr., Chinn, M. D., and Diehl, R. D.: Phys. Rev. B (1980) $\underline{21}$, 4170 and references contained therein.

79) Rauber, S., Klein, J. R., Cole, M. W., and Bruch, L. W.: Surface Sci. (1982) $\underline{123}$, 173.

80) Suzanne, J., Coulomb, J. P., and Bienfait, M.: Surface Sci. (1975) $\underline{47}$, 204.

81) Singleton, J. H. and Halsey, G. D. Jr.: Can. J. Chem. (1955) $\underline{33}$, 184.

82) De Oliveira, M. J. and Griffiths, R. B.: Surface Sci. (1978) $\underline{71}$, 687.

83) Larher, Y. and Terlain, A.: J. Chem. Phys. (1980) $\underline{72}$, 1052 and Terlain, A. and Larher, Y.: Surface Sci. (1980) $\underline{93}$, 64.

84) Tessier, C.: Thèse d'état, Paris (1984) (unpublished).

85) Polanco, S. E. and Bretz, M.: Phys. Rev. B (1978) $\underline{17}$, 151.

86) Pierre, L., Guignes, H., and Lhuillier, C.: (to be published).

87) Richards, M. G.: (private communication).

METALS IN INTIMATE CONTACT

John R. Smith
Physics Department, General Motors Research
Warren, MI 48090-9055

John Ferrante
National Aeronautics and Space Administration
Lewis Research Center
Cleveland, OH 44135

ABSTRACT

We consider the situation where two metal surfaces are in such close contact that electron wave functions from the two metals can overlap. The electronic structure and total energy is computed as a function of separation between the surfaces for all combinations of the simple metals Al(111), Zn (0001), Mg(0001), and Na(110). Significant electron transfer and/or rearrangement was found. The range of relatively strong bonding was about one interplanar spacing. The dominant attractive energy component is the electron exchange-correlation energy. The kinetic energy component is the dominant repulsive component at small separations but becomes attractive at larger separations. Several analogies are drawn between the bimetallic and molecular bond. In fact, a universality is discovered for the shape of the total energy as a function of distance between atoms. Finally, the bimetallic interface between simple metals is compared with that between transition metals. The transition metal localized orbitals lead to a more localized electronic rearrangement at the interface.

INTRODUCTION

Metal surfaces in intimate contact offer interesting scientific questions and at the same time important technological ones. These interfaces are involved in the processes of friction and wear, deposition of metal films, fracture, grain boundary energetics, and electrical contact phenomena [1].

Relatively strong bonds can be formed between such metal surfaces when spacings become as small as a few angstroms. In that case, electron wave functions from the two metals can overlap. This can lead to large electron-exchange interactions and -- when the metals are not identical -- to charge transfer. Total energies tend to vary relatively rapidly with spacing between the surfaces. Similarly rapid variation is found in electron density distributions in the interface.

With such electronic rearrangement, one cannot avoid fully self-consistent calculations. That is, the electron wave functions must be consistent with the potential used to compute them. It is this requirement, plus the loss of symmetry in the direction perpendicular to the interface, which are the main reasons why there is so little theoretical work on bimetallic interfaces.

The first quantum-mechanical treatment of the bimetallic interface was done by Bennett and Duke [2]. Binding energies were not computed and the calculations were not fully self-consistent. Binding energies were later computed by the authors [3] as a function of distance between the metal surfaces. The electron density was approximated in Ref. 3 as a simple overlap of solid-vacuum solutions, so the calculations were not self-consistent. Fully self-consistent computations of interfacial electron density distributions and binding energies were next determined by the authors for adhesive interactions between identical metal surfaces of Al, Mg, Zn, and Na. These calculations were carried out over a range of separations so that the shape of the binding energy curve was determined. Using quite different methods appropriate to d-band metals, Richter, Smith and Gay [4] (see also Ref. [5]), carried out the first self-consistent calculation of the surface energies of a series of transition metals.

In the following we will describe the first self-consistent treatment of adhesive energy curves and electronic structure for bimetallic interfaces. All combinations of Al(111), Zn(0001), Mg(0001), and Na(110) will be considered. What are some of the things we should be looking for in the results? We will attempt to determine the dominant energy component contributing to the bimetallic bond. The range or distance over which the bimetallic interaction is relatively strong would also be of interest. We will be looking for (and finding) relationships between binding energy curves for bimetallic interfaces and those for diatomic molecules. In fact, these bimetallic calculations led to the discovery [6] of a universal binding energy relation which was later found to extend to chemisorption [7], cohesion and diatomic molecule energetics [8], and even to nuclear matter [9]. On the electronic structure, we will attempt to ascertain any relationship between contact potentials and electronic barriers in the interface. Finally, we will compare electronic charge rearrangements found at transition metal interfaces with those at simple metal interfaces.

Theory of the symple Metal Interface

As this is a solid state theory conference, it is perhaps appropriate to dwell on theoretical techniques. However, much of the techniques for simple, bimetallic interfaces are similar to that for interfaces between identical, simple metals. The latter were discussed at length by the authors [10] earlier. Also details of theoretical methods for the simple, bimetallic interfaces can be found in Ref. [11].

So we take the opportunity now to discuss our approach in a somewhat pedagogical fashion. We approach the problem in two steps. First, we solve the Kohn-Sham [12] equations self-consistently for a bi-jellium model. The bi-jellium model is shown in Fig. 1 for the Al(111) - Mg(0001) interface. We have two jellia--uniform positive density backgrounds -- separated by a distance a. This nearly-free-electron ansatz is of course only valid for simple metals. By charge neutrality, a = 0.0 when the distance between Al and Mg atomic planes is $\frac{d_{Al} + d_{Mg}}{2}$, where d_{Al} and d_{Mg} are distances between planes of atoms in the two metals. The Kohn-Sham equations for the electron density in this model are

$$\left(-\frac{1}{2}\frac{d^2}{dy^2} + v_{eff}(n,y)\right) \psi_k^{(i)}(y) = \frac{1}{2}(k^2 - k_F^2) \psi_k^{(i)}(y) \quad , \tag{1}$$

where k is the Bloch vector component in the y-direction (Fig. 1) and

$$v_{eff}(n;y) = \phi(y,a) + \frac{\delta E_{xc}\{n(y,a)\}}{\delta n(y,a)} \quad , \tag{2}$$

The electrostatic potential $\phi(y,a)$ satisifies Poisson's equation

$$\frac{d^2\phi(y,a)}{dy^2} = -4\pi [n(y,a) - n_+(y)] \quad . \tag{3}$$

$n_+(y)$ is the jellium density and $n(y,a)$ is the electron density (Fig. 1);

$$n(y,a) = \sum |\psi_k^{(i)}(y)|^2 \tag{4}$$

and the sum is over all occupied states (including k_x and k_z in the $e^{ik_x x}$ and $e^{ik_z z}$ components of the total wave function). $E_{xc}\{n(y,a)\}$ is the electron exchange-correlation energy. The simultaneous solution of Eq. 1-4 is what is meant by a self-consistent solution. Because of the form of Eq. 1 and 2, this many-electron problem has been transformed into an effective one-electron one which is valid as far as the computation of the total energy [12]. This and the bi-jellium model makes the problem tractable.

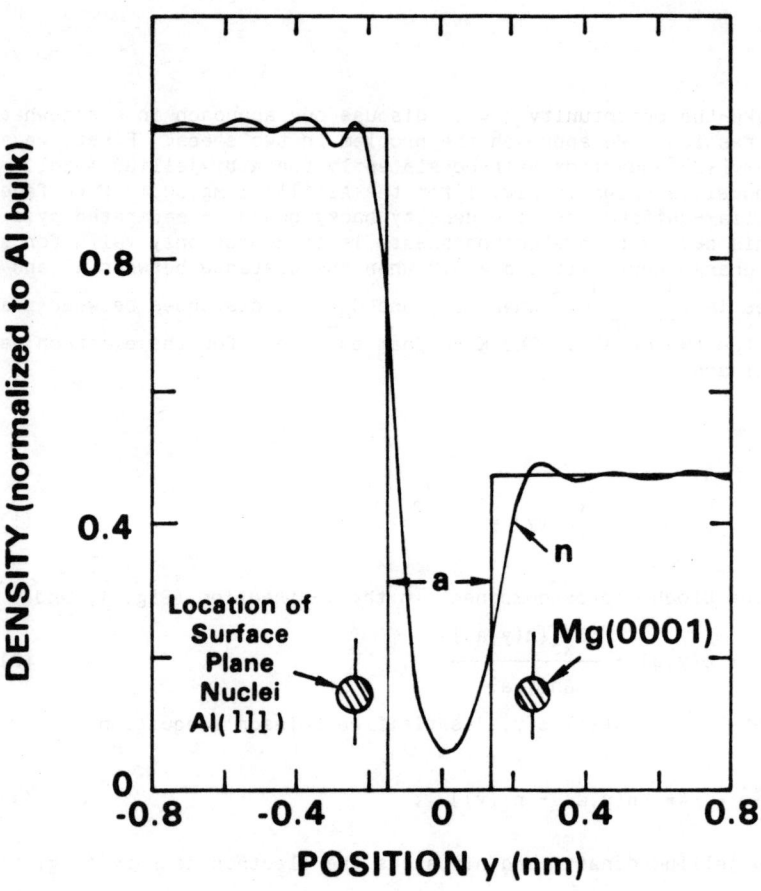

Figure 1. Electron number density n and jellium ion charge density n_+ for an Al-Mg contact. When a = 0.0, the distance between Mg and Al atomic planes is $(d_{Al} + d_{Mg})/2$, where d_{Al} and d_{Mg} are the respective bulk interplanar spacings.

The second of the two steps involves the inclusion of the crystal structure of the two metals in first-order perturbation theory. That is,

$$E\{n(\vec{r})\} = E\{n(y,a)\} + A\int \delta v(y,a)n(y,a)dy \tag{5}$$

where $\delta v(y,a)$ is the average, over planes parallel to the surface, of the difference in potential due to an array of pseudopotentials and that from a jellium surface and A is the cross-sectional area. Eq. 5 is a good approximation for the closest-packed plane of simple metals like Al, Zn, Mg and Na because $\delta v(y,a)$ is relatively small for them.

The adhesive interaction energy, E_{ad}, between two metals separated by the distance a is now calculable:

$$E_{ad} = [E(a) - E(\infty)]/2A. \tag{6}$$

The theoretical approach to the simple metal interface has now been stated in Eq. 1-6. The details of the solutions of these equations we leave to Ref. 10-11. The results are of more interest here.

RESULTS AND DISCUSSION

Results from Eq. 1-4 for an example interface Al-Mg are shown in Fig. 2 for three separations, $a = 0.0$, 0.16, and 1.6 nm (0, 3, and 30 a.u.). This shows that the self-consistent electron density distributions are very sensitive to small changes in interfacial separation. This sensitivity is further exemplified in Fig. 3 where the self-consistent, effective-one-electron potential energy functions, $v_{eff}(y)$, are plotted for Al-Mg at the same three separations as in Fig. 2. Even at $a = 0.0$, there is a potential step leading to nonzero electron reflection coefficients at the interface. Out of this step a barrier rises rapidly with increasing a. The computed contact potential between these two metals is only 0.21 eV, and it is independent of a as long as there is sufficient wave function overlap between the two metals so that the Fermi levels will equilibrate due to charge transfer. The contact potential is determined from the difference between computed electron work functions of the two metals.

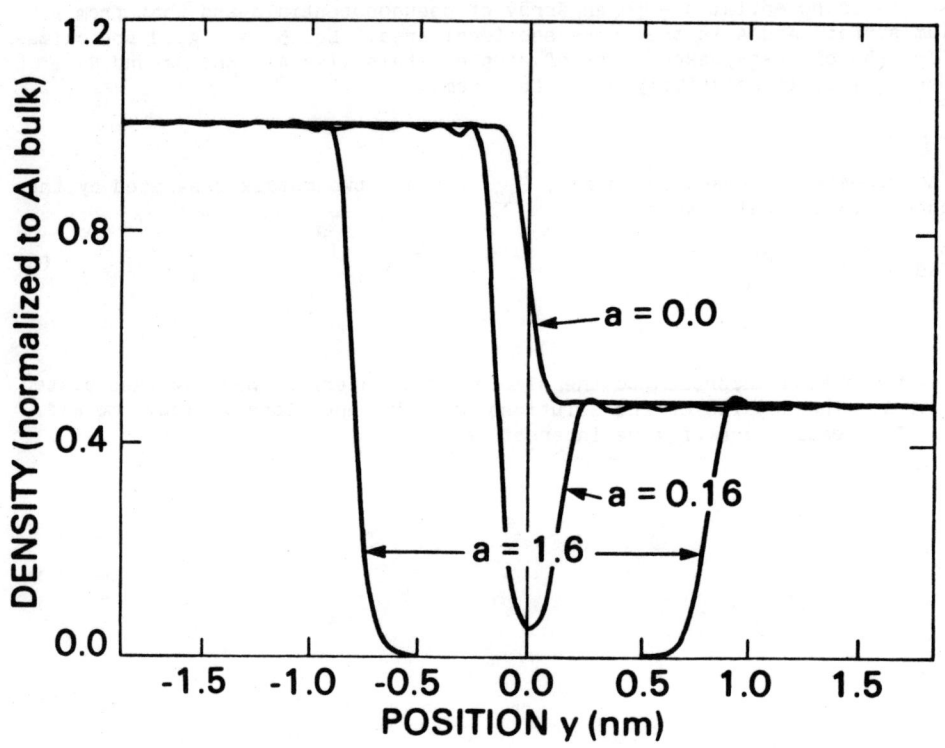

Figure 2. Electron density versus position for an Al-Mg contact (Al is on the left), for separations of 0, 0.16 and 1.6 nm.

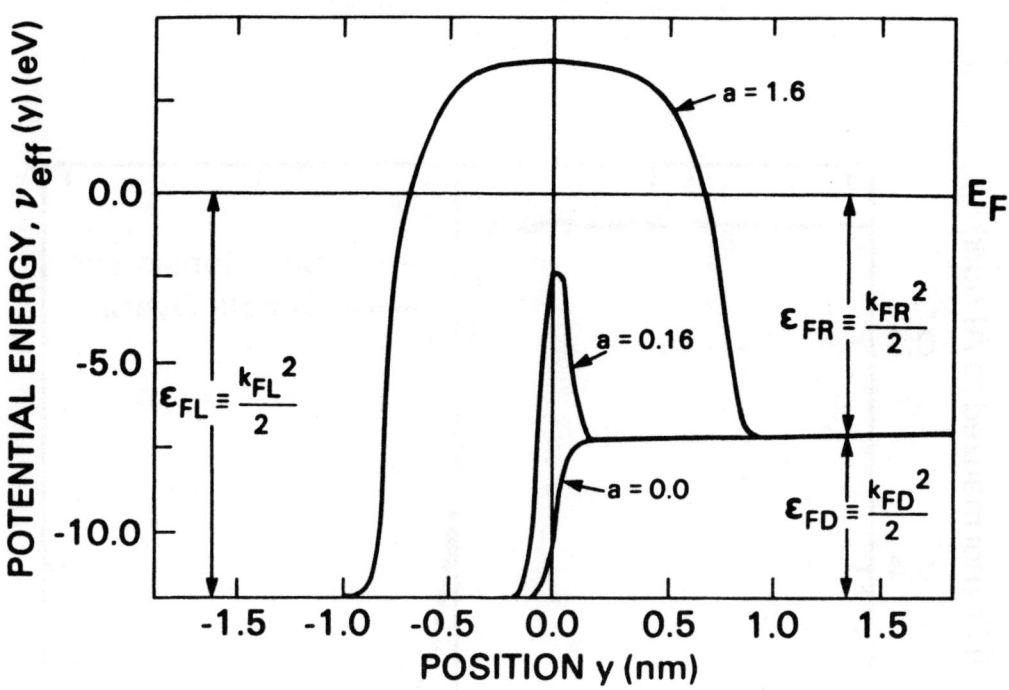

Figure 3. Electron potential energy $\nu_{eff}(y)$ vs position for an Al-Mg contact (Al is on the left), at separations of 0, 0.16, and 1.6 nm.

Speaking of charge transfer, Fig. 4 exhibits the self-consistent electron density distribution, $n(y,0)$, for the Al-Na contact at $a = 0.0$. For comparison, the result of linearly adding the solid-vacuum solutions is also plotted. The two density distributions are quite similar. The difference (self-consistent minus simple overlap), is due to the charge transfer and/or rearrangement when the contact is formed. While this difference is small, it is significant as shown in Fig. 5. In Fig. 5 a), the density difference is plotted and 5 b) is the corresponding difference in $\nu_{eff}(y,0)$. Note the large peak in the potential, indicating the significance of the charge transfer/-rearrangement. This peak is much larger than the contact potential ΔV, as shown. It is not obvious from these plots at zero separation that there is a net electron transfer from the Na to the Al, as there must be since the Al work function is larger than that of the Na. The same difference plots at $a = 1.59$ nm = 30 au (Fig. 6) clearly shows this to be the case. At this larger separation, there is a region of space containing very little charge, which is the reason the potential difference falls off linearly in that region. In this case, we have a text-book example of a net charge transfer leading to a contact potential barrier. Note that ΔV in Fig. 5 is the same as ΔV in Fig. 4, as it must be.

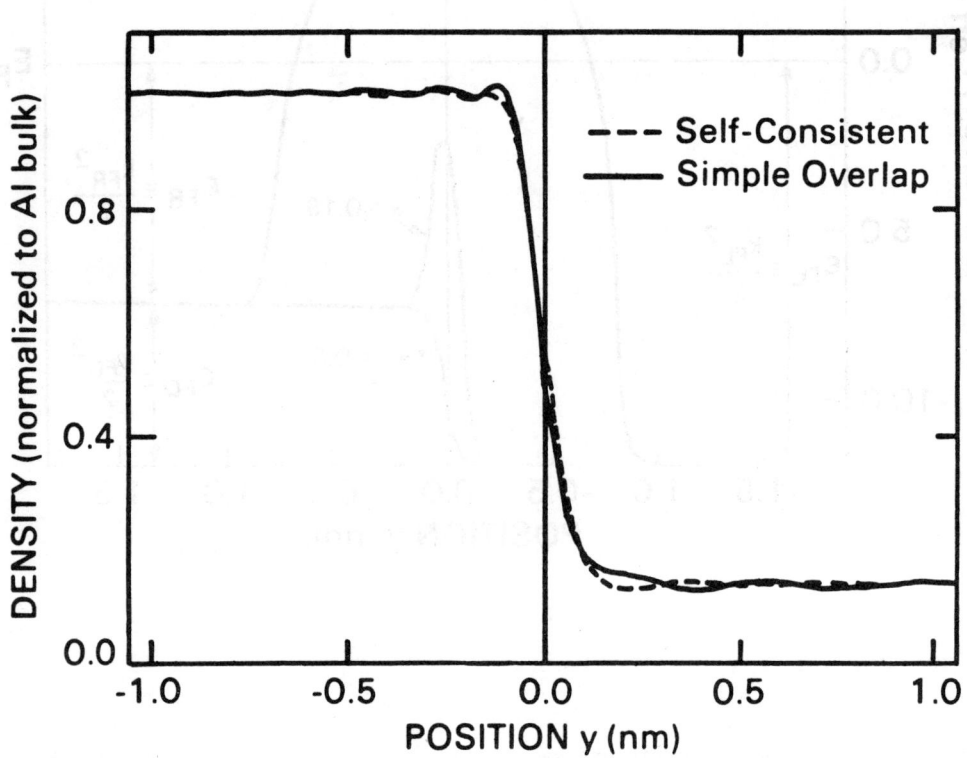

Figure 4. Electron density distributions in the Al-Na interface at zero separation (a = 0.0). The dashed curve is the fully self-consistent result, while the solid curve results from a simple overlap of the solid-vacuum results for each of the two metals.

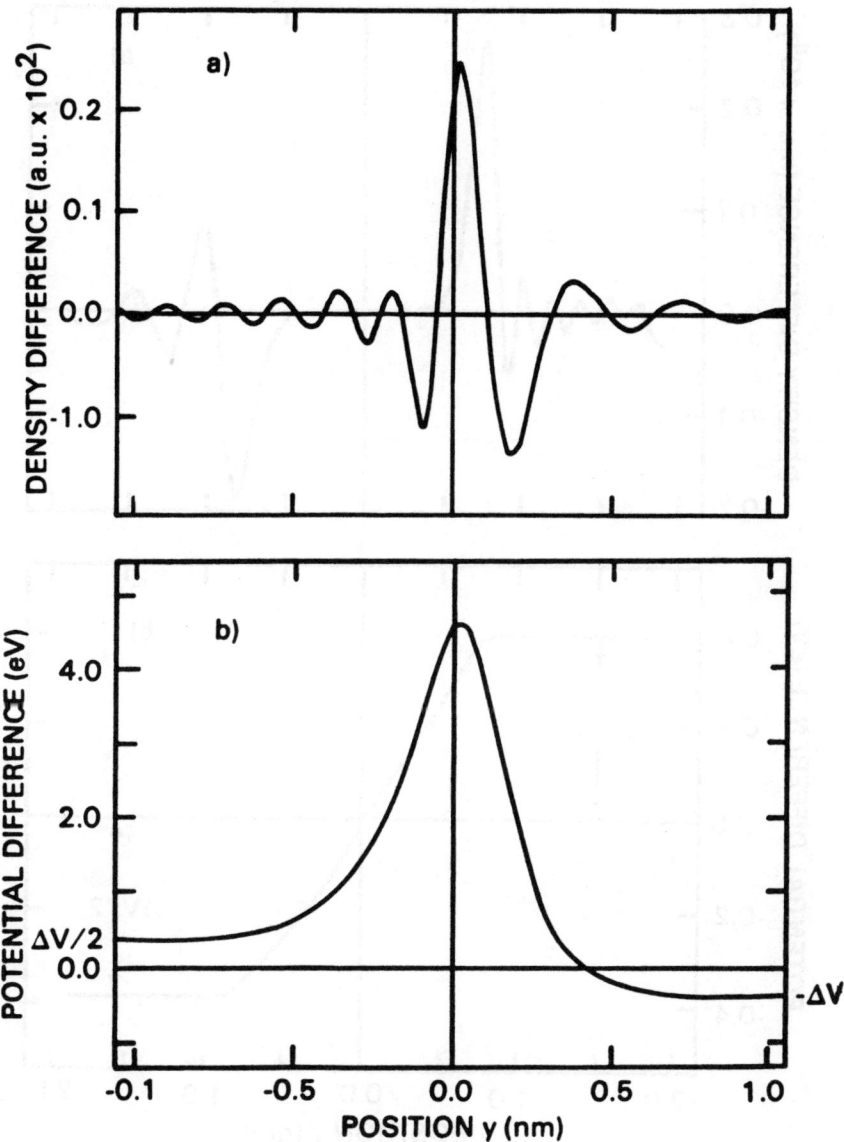

Figure 5. a) Self-consistent electron densities, $n(y,a)$, minus overlapped-solid-vacuum electron densities for $a = 0.0$ in an Al-Na contact, (Al on the left). This density difference plot shows the charge rearrangement due to interaction between the Al and Na surfaces.

b) Self-consistent electron potentials $v_{eff}(n;y)$, minus overlapped-solid-vacuum electron potentials, for $a = 0.0$ in an Al-Na contact. ΔV is the contact potential.

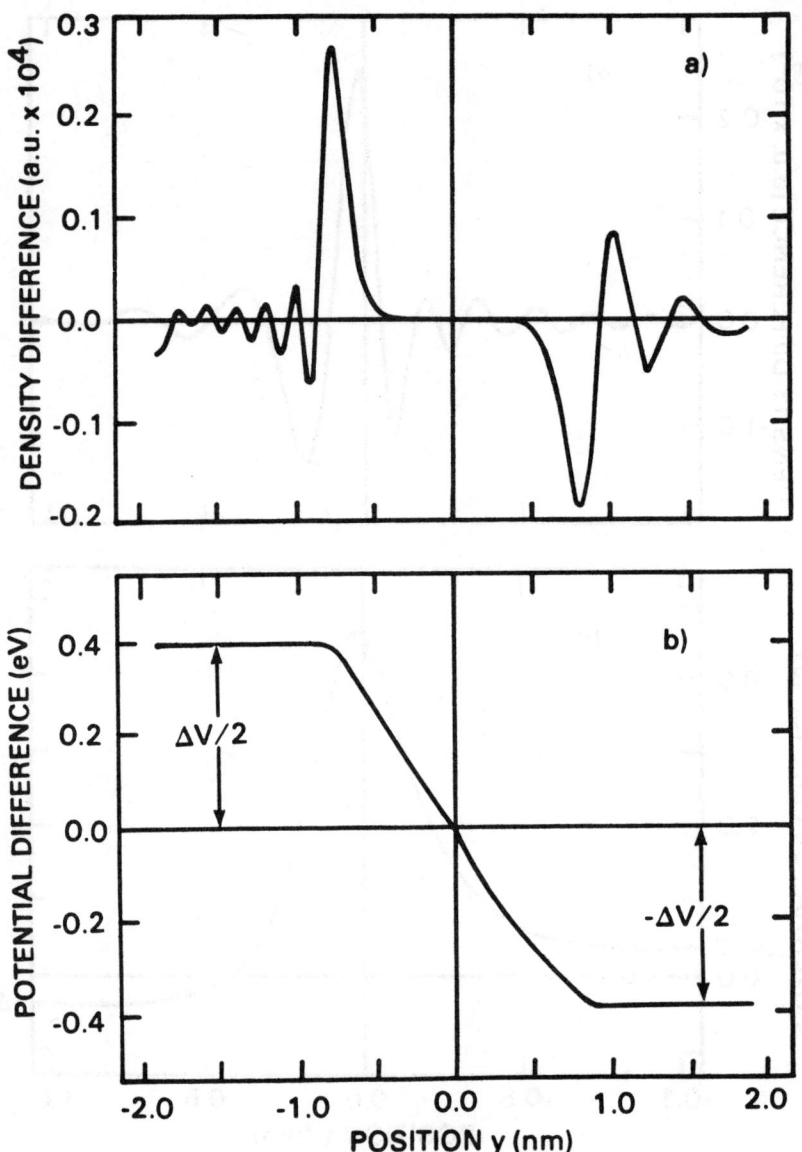

Figure 6. a) Self-consistent electron densities, n(y,a), minus overlapped-solid-vacuum densities, n(y,a), for a = 1.59 nm in an Al-Na contact, Al on the left. This shows clearly that electrons are transferred from the Na to the Al.
b) Self-consistent potentials, v_{eff}(n;y), minus overlapped solid-vacuum electron potentials for a = 1.59 nm in an Al-Na contact. ΔV is the contact potential resulting from the charge transfer shown in Figure 6 a).

Now let us compute the total energies, Eq. 5. In this initial calculation for bimetallic interfaces, we ignore any distortions like dislocations which can occur in contacts when the two metals are not identical [13]. The results for E_{ad} are shown in Fig. 7. First note that the range of strong bonding, i.e. the range over which the adhesive force is relatively strong is about 0.2 nm, roughly a bulk interplanar spacing. Secondly, the shapes and well depths of the plots in Fig. 7 cover a broad range. It was discovered, nevertheless, that the following simple scaling of the results of Fig. 7 and our earlier results on identical metal contacts [10] leads to a universal energy relation for adhesion. The deviation of the separation a from the equilibrium value a_m is scaled as

$$a^* \equiv \frac{(a-a_m)}{\ell} \tag{7}$$

and the ordinate $E_{ad}(a)$ is scaled as

$$E^*_{ad}(a) \equiv E_{ad}(a)/\Delta E \tag{8}$$

where ΔE is the magnitude of $E_{ad}(a_m)$ (for identical metal contacts, ΔE is the surface energy). The scaling length ℓ is given by

$$\ell = \left\{ \Delta E \left[\frac{d^2 E(a)}{da^2} \right]^{-1}_{a_m} \right\}^{1/2} \tag{9}$$

so that

$$\left[\frac{d^2 E^*_{ad}(a^*)}{da^{*2}} \right]_0 = 1. \tag{10}$$

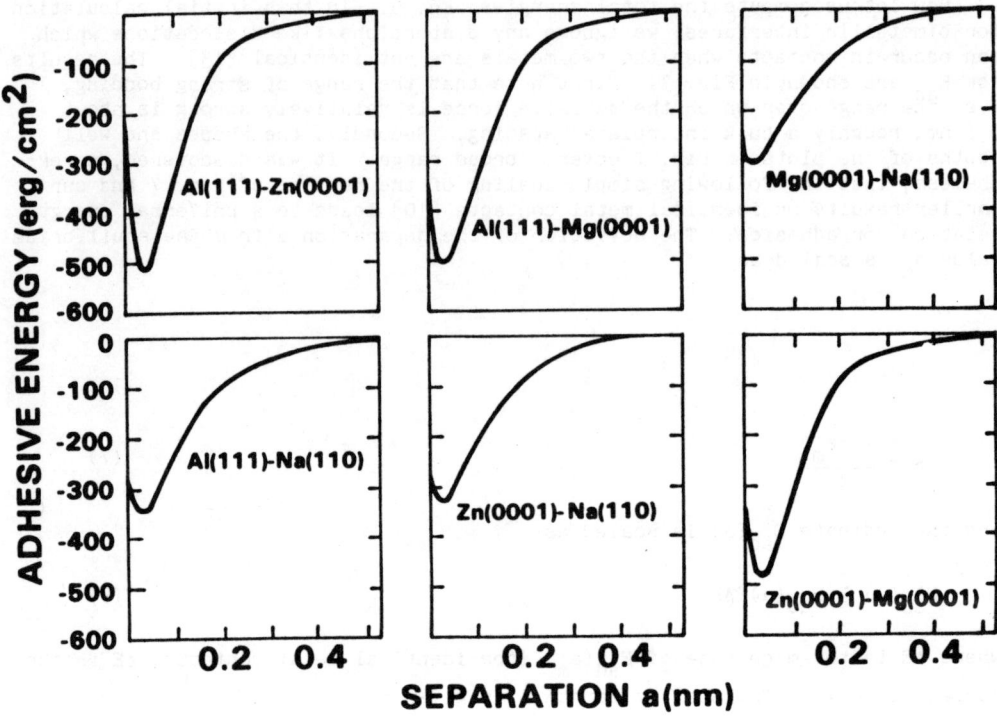

Figure 7. Adhesive binding energy versus separation a. Incommensurate adhesion is assumed.

The results of scaling the curves of Fig. 7 and of Ref. 10 are shown in Fig. 8. One can see that the $E_{ad}(a)$ for all ten contacts fall very closely on a single, universal curve. This means we could have, for example, solved the Al-Al contact and would then have known the shape of $E_{ad}(a)$ for all nine other contacts. This would have been a very significant savings because self-consistent total energy calculations at defects - such as we have described here - are difficult and must be done for each separation for each contact. They have been made possible only in the last 2-3 years with modern computers. So "Mother Nature" has truly smiled on us here. In fact, this universality extends to chemisorption [7, 14], cohesion of metals, and certain diatomic molecule energetics [8] as shown for representative cases in Fig. 9.

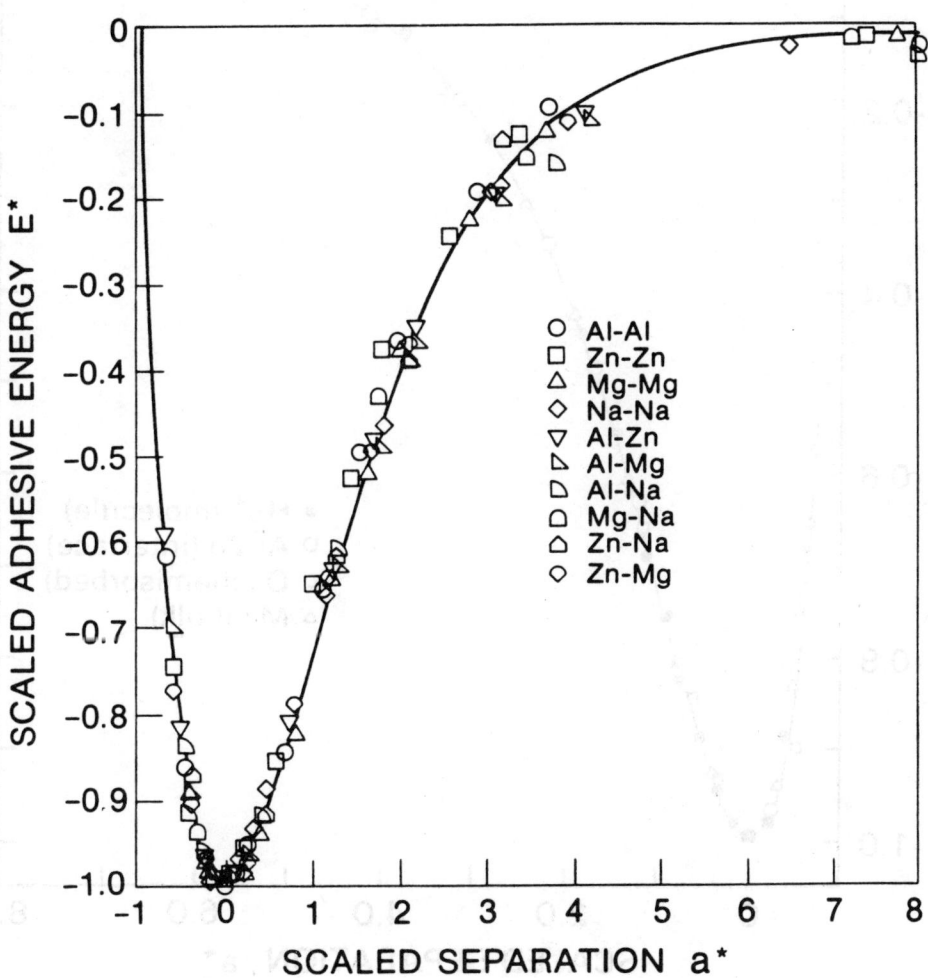

Figure 8. Adhesive energy results from Figure 7 and Ref. 10 scaled as described in the text (see Eq. 7-8).

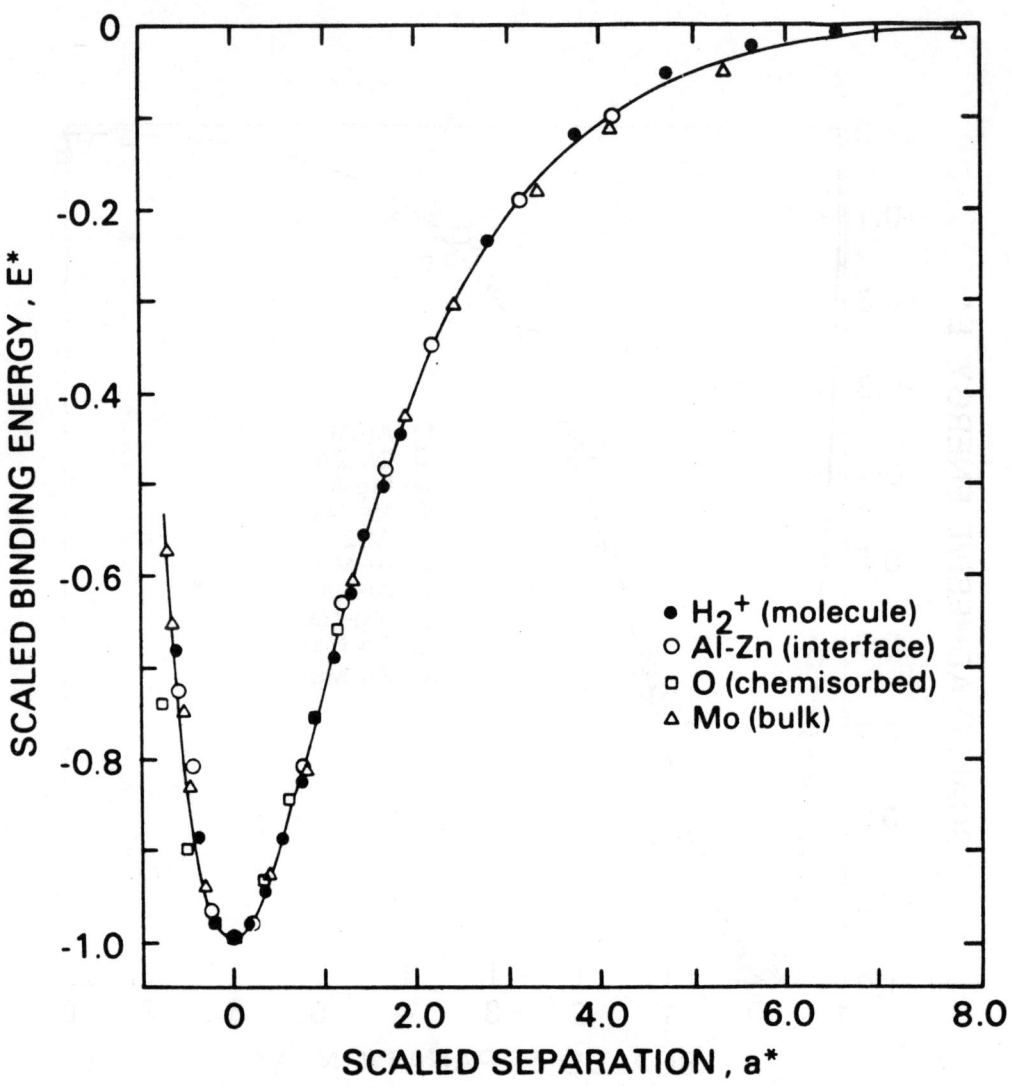

Figure 9. Binding energy as a function of interatomic separation for four systems as noted, scaled as described in the text. See Ref. [8].

We have further evidence of a connection between bimetallic and molecular interactions. The components of the total energy are plotted in Fig. 10 for the Al(111) - Mg(0001) contact. This qualitative behavior is typical. At large separations (>0.2 nm) the kinetic energy component is negative relative to infinite separation. That is, the kinetic energy "initiates" the bond, presumably due to smoothing of the electron wave functions in the bonding region. The electrostatic energy is positive at these same relatively large separations but changes sign at smaller separations. Note that the dominant attractive energy component is the exchange-correlation energy, while the dominant repulsive term at small separations is the kinetic energy. This behavior of the kinetic and electrostatic energy components was discovered earlier for diatomic molecules [15] and it is qualitatively identical to what we have found for the bimetallic interface (Fig. 10). Earlier we [10] reported the same behavior for contacts between identical metals.

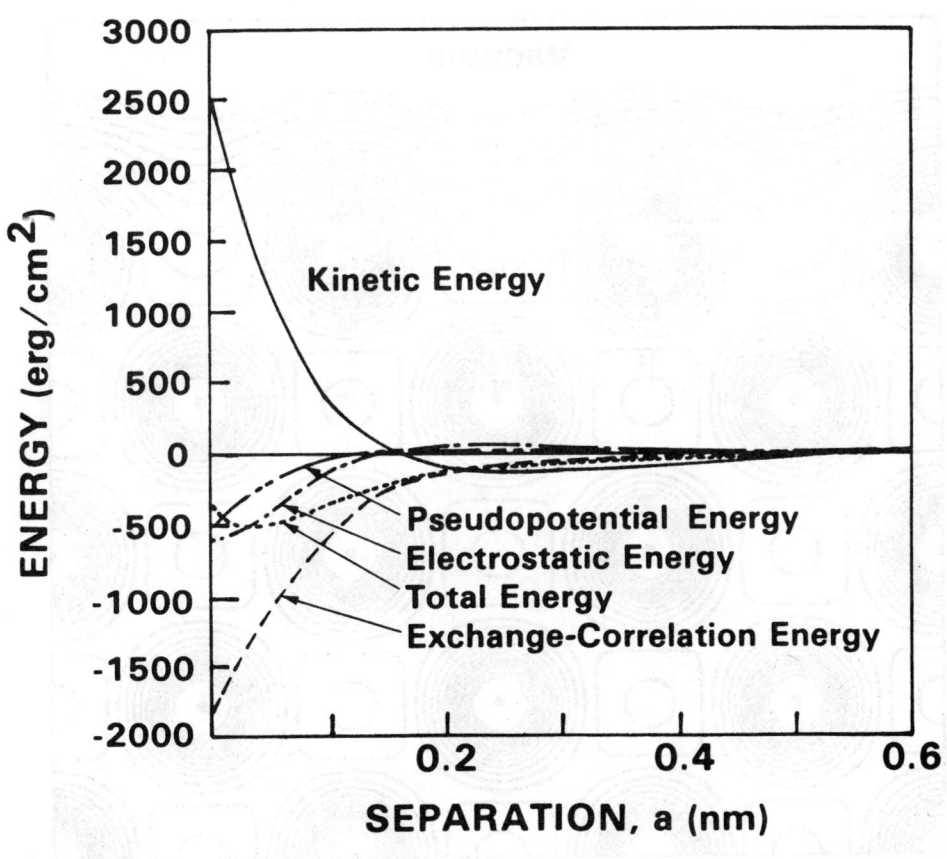

Figure 10. Self-consistent energy components of the binding energy for an Al(111)-Mg(0001) contact.

Recently we have found [9] evidence that the universality may apply also to <u>nuclear matter</u> and electron-hole liquids. The latter can be found in semi-conductors like Si and Ge.

Comparison with the Transition Metal Interface

Interfaces between metals with d-bands present a different kind of problem than the simple metal interfaces described above. Because of the localized nature of the d-electrons, the electron density can vary by a factor of 10 in a distance of about 1 angstrom as shown in the self-consistent charge contours of Fig. 11, taken from Ref. [16]. These exhibit the electron density distribution near a Ni(100) surface. As in the case of the simple metals, there is a charge rearrangement at the surface, but in this case it is a localized rearrangement (Fig. 11). One could not conceive of modelling transition metal interfaces with a bijellium, perturbation-theory approach as we used for simple metal interfaces.

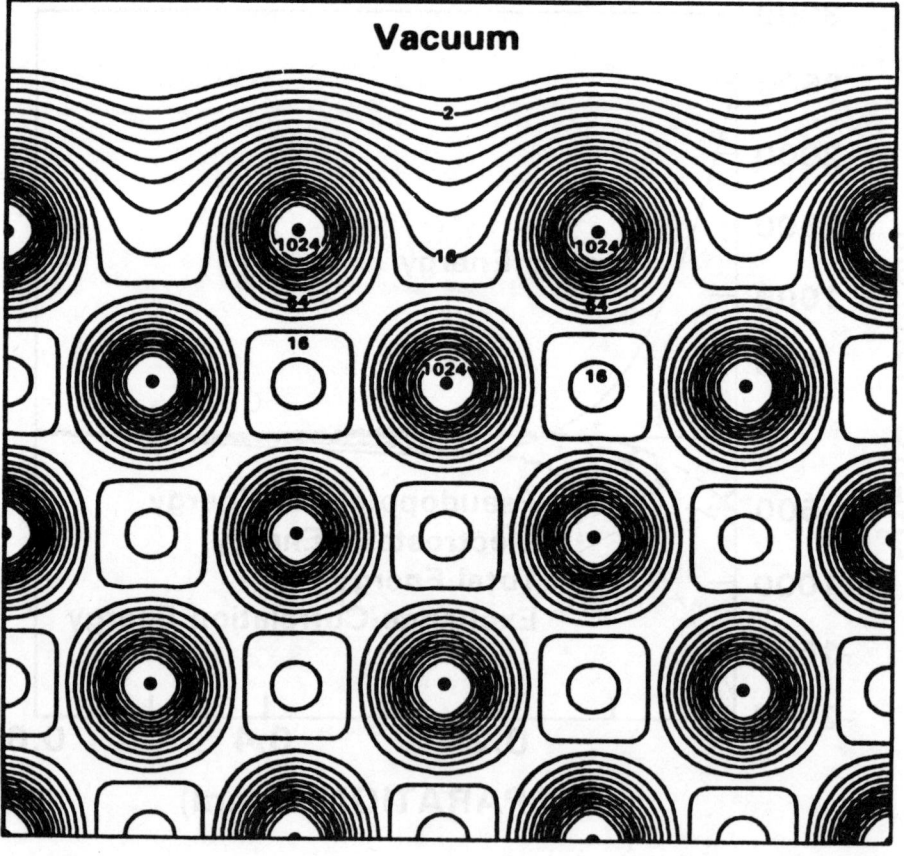

Figure 11. Electronic charge density contours at a Ni(100) surface, from Ref. [16].

A method based on localized basis functions appropriate for transition metals - the Self-Consistent Local Orbital Method - is described at length in Ref. 17. An example result [18] is shown in Fig. 12 for the Ag-Pd(100) interface. Here a monolayer of Ag was deposited on a Pd(100) surface epitaxially. This is a convenient configuration experimentally because electrons can be photoemitted through the thin Ag layer to obtain information about the interface [18]. Given the short screening lengths metals have, even a monolayer of Ag is representative of a thicker Ag film interface. In Fig. 12, the sum of the isolated Ag monolayer electron density distribution and the clean Pd(100) surface distribution is subtracted from the distribution of the Ag monolayer chemisorbed on the Pd(100). This difference then is due to the charge transfer and/or rearrangement occurring upon interaction between the Ag and Pd(100). It is the transition metal analogue to Fig. 4a. Dashed contours refer to a charge depletion and solid contours to an accumulation. Here we see that electrons have been "scooped up" from around both the Ag and Pd atoms and deposited in the interstitial region between the Ag and Pd layers. This is not unlike what one finds in a covalently-bonded molecule, so as in simple bimetallic interfaces we find a molecular analogy.

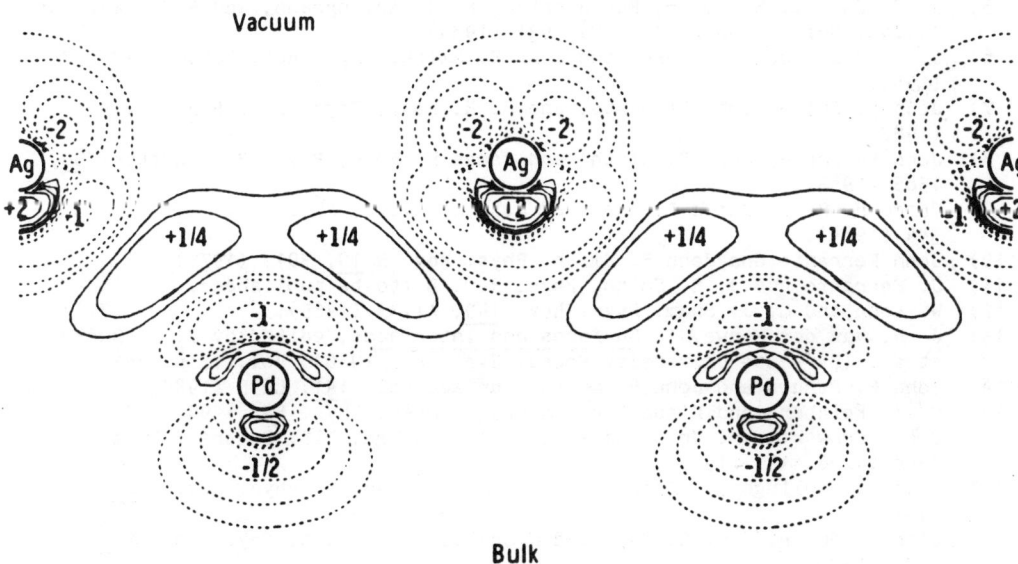

Figure 12. Charge transfer of the palladium {100} slab upon silver adsorption. The difference density $\rho(Pd\{100\} + \rho(1\times 1)Ag)_3 - \rho(Pd\{100\}) - \rho(Ag)$ is plotted along a (110) plane in units of 10^{-2} e/a$_o^3$. Successive contours are separated by a factor of two. Solid (dashed) lines denote an accretion (depletion) of electrons after adsorption.

Very recently a method to compute total energies for transition metal interfaces has been formulated [4]. It is of course very different from the simple-metal method described earlier. With it, transition metal surface energies have been computed for the first time, in good agreement with experiment [4]. It was found that a highly accurate Hamiltonian was required to make the calculations viable. There was a suggestion in the results of a surface energy correlation with d-band filling.

ACKNOWLEDGEMENTS

The authors are grateful to Jack Gay, Jim Burkstrand, Jim Rose, and Roy Richter for useful discussions, and especially to Jan Herbst for his comments regarding numerical techniques.

REFERENCES

1) D. H. Buckley, J. Colloid and Interface Sci. $\underline{58}$, 36 (1977).
2) Alan J. Bennett and C. B. Duke, Phys. Rev. $\underline{160}$, 541 (1967); $\underline{162}$, 578 (1967).
3) John Ferrante and John R. Smith, Surface Sci. $\underline{38}$, 77 (1973).
4) Roy Richter, J. R. Smith, and J. G. Gay, Proceedings of the First International Conference on the Structure of Surfaces, Berkeley, August, 1984, Springer-Verlag, edited by M. A. Van Hove and S. Y. Tong.
5) J. G. Gay, J. R. Smith, Roy Richter, F. J. Arlinghaus, and R. H. Wagoner, J. Vac. Sci. Technol. A $\underline{2}$ (2), 931 (1984).
6) J. H. Rose, John Ferrante and John R. Smith, Phys. Rev. Letters $\underline{47}$, 675 (1981).
7) John R. Smith, John Ferrante, and J. H. Rose, Phys. Rev. B $\underline{25}$, 1419 (1982).
8) John Ferrante, John R. Smith, and James H. Rose, Phys. Rev. Letters $\underline{50}$, 1385 (1983).
9) James H. Rose, James P. Vary and John R. Smith, Phys. Rev. Letters $\underline{53}$, 344 (1984).
10) John Ferrante and John R. Smith, Phys. Rev. B $\underline{19}$, 3911 (1979).
11) J. Ferrante and J. R. Smith, Phys. Rev. B (to be published).
12) W. Kohn and L. J. Sham, Phys. Rev. $\underline{140}$, A1133 (1965).
13) J. H. van der Merwe, in <u>Surfaces and Interfaces</u>, edited by J. J. Burke, et al. (Syracuse University Press, Syracuse, 1966), Vol. 1, p. 361.
14) John P. Perdew and John R. Smith, Surface Sci. $\underline{141}$, L295 (1984).
15) M. J. Feinberg and Klaus Ruedenberg, J. Chem. Phys. $\underline{54}$, 1495 (1971). C. Woodrow Wilson, Jr., and William A. Goddard, III, Theoret. Chem. Acta (Berl.) $\underline{26}$, 195 (1972).
16) Frank J. Arlinghaus, Jack G. Gay, and J. R. Smith, Phys. Rev. B $\underline{21}$, 2055 (1980).
17) John R. Smith, Jack G. Gay, and Frank J. Arlinghaus, Phys. Rev. B $\underline{21}$, 2201 (1980).
18) T. W. Capehart, R. Richter, J. G. Gay, J. R. Smith, J. C. Buchholz, and F. J. Arlinghaus, J. Vac. Sci. and Techn. $\underline{A1}$, 1214 (1983).
19) F. J. Arlinghaus, J. G. Gay, and J. R. Smtih, Phys. Rev. B $\underline{21}$, 2055 (1980).

CHEMICAL TRENDS OF SCHOTTKY BARRIERS

John D. Dow
Department of Physics, University of Notre Dame
Notre Dame, Indiana 46556

and

Otto F. Sankey
Department of Physics, Arizona State University
Tempe, Arizona 85287

and

Roland E. Allen
Department of Physics, Texas A&M University
College Station, Texas 77843

ABSTRACT

The observed chemical trends in Schottky barrier heights (i.e., the variation in the barrier height ϕ_B as a function of the alloy composition x, or the dependence of the barrier height on the anion or cation species) are explained by Fermi-level pinning due to defects. Microscopic calculations of surface defect levels, rather than phenomenological arguments, are presented to support this viewpoint. We find that the slope of the pinning defect level as a function of alloy composition (dE/dx) is a signature of the defect type. In the case of $Al_xGa_{1-x}As/Au$ contacts for all compositions x, and for $Al_xGa_{1-x}As/Al$ and $Al_xGa_{1-x}As/In$ contacts for large x, the Schottky barrier heights are attributed to Fermi-level pinning by cation-on-anion-site antisite defects (|dE/dx| is large). $Al_xGa_{1-x}As/Al$ and $Al_xGa_{1-x}As/In$ Schottky barriers, for small x, are attributed to Fermi-level pinning by anion-on-the-cation-site antisite defects (|dE/dx| is small). This interpretation is supported by both detailed calculations and the results of a simple four-atom model.

I. INTRODUCTION

Over the years, there have been many attempts to understand the observed chemical trends in Schottky barrier heights ϕ_B -- the dependence of ϕ_B on the anion or cation species, or on the alloy composition x. In the past, such attempts have had the disadvantage that no fundamental microscopic foundation has been available. The introduction of Bardeen's concept of Fermi-level pinning [1] and Spicer's defect model [2-5], however, have provided a general framework that makes it possible to understand chemical trends in ϕ_B from a microscopic point of view: Since the Schottky barrier height in the Fermi-level pinning model is approximately equal to the difference between a band edge (conduction band edge for an n-type semiconductor and valence band edge for p-type) and the relevant defect level (lowest acceptor level for n-type; highest donor level for p-type), chemical trends in barrier heights are explained by the combined chemical trends in band edges and "deep" defect levels at the semiconductor/metal contact.

Recently we have reported theoretical predictions of Schottky barrier heights for Au contacts to various III-V alloys [6] and for transition-metal contacts to Si_xGe_{1-x} alloys [7]. The III-V/Au barriers are attributed to Fermi-level pinning by cation-on-anion-site III-V surface antisite defects [6]. The Si_xGe_{1-x} barriers are attributed to Fermi-level pinning by interfacial dangling bonds [7]. For both systems, the theory is in quite good agreement with the measured barrier heights, with the observed chemical trends being particularly well described by the theory.

Here we extend the simple theory of Ref. [6] and consider both antisite defects, including the the anion-on-cation-site antisite defect. Our principal motivation is to compare the alloy dependences for the Schottky barriers that result from Fermi-level pinning by the two different types of antisite defects. As discussed below, we find that dE/dx is very different for the two defects in some cases, where E is a Fermi-level pinning defect energy level and x is the alloy composition. This appears to explain the different dependences on x of observed barrier heights of $Al_xGa_{1-x}As$ with Al and In contacts on the one hand, and with Au contacts on the other hand.

II. Simple Four-Atom Model

Before giving the results of our detailed calculations -- which employ the sp^3s* model of Vogl et al. [8] for the bulk electronic structure, the scaled-atomic energy model of Hjalmarson et al. [9] for the impurity potentials, and the analytic Green's function technique [10] -- let us consider a very simple four-atom model for each of the two surface antisite defects: the antisite atom at a surface and its three nearest-neighbors. We will find that this model provides a remarkably good description of the chemical trends, and tends to increase our faith in the central results of the much more complicated calculations.

The simple four-atom model can be constructed by first considering a five-atom model consisting of an antisite impurity in the bulk and its four neighbors, and then replacing one of the four neighboring atoms by a vacancy -- to simulate the semiconductor surface. In the bulk, an anion or cation antisite defect is tetrahedrally coordinated, which leads to deep level electronic states of A_1 (s-like) or T_2 (p-like) symmetry. The symmetry is reduced at the surface, and the states of A_1 and T_2 symmetry mix.

A. Bulk antisite Defects

For concreteness, consider the anion-site bulk antisite defect (Ga_{As}) in GaAs. Take as a basis (i) the s- and p-orbitals of the antisite defect atoms $|s\rangle$ and $|p\rangle$ (with energies ε_s and ε_p) and (ii) the main s-like (or A_1-symmetric) and p-like (T_2-symmetric) orbitals of the rest of the solid without the central atom -- namely the A_1 and T_2 orbitals of a vacancy (with energies $E(A_1;v)$ and $E(T_2;v)$). In a model which considers only the defect and its four neighbors, the vacancy A_1 orbital is

$$|A_1;v\rangle = (\ |1\rangle + |2\rangle + |3\rangle + |4\rangle\ )/2, \tag{1}$$

where $|i\rangle$ is the inward-directed sp^3-hybrid centered on the i-th neighboring site [11]. Similarly the relevant T_2-vacancy orbital is

$$|T_2;v\rangle = (12)^{-1/2}\ (\ |1\rangle + |2\rangle + |3\rangle - 3\ |4\rangle\ ). \tag{2}$$

The s orbital of the antisite impurity only interacts with $|A_1;v\rangle$; and the p orbital which is polarized toward atom 4 interacts only with $|T_2;v\rangle$. Notice that the wavefunction is equally distributed among the four hybrids for the $|A_1;v\rangle$ orbital, but is more heavily weighted on hybrid $|4\rangle$ for the $|T_2;v\rangle$ orbital. The model **bulk** antisite Hamiltonian can be simply written as a direct sum:

$$H_{bulk} = \begin{pmatrix} H(A_1) & 0 \\ 0 & H(T_2) \end{pmatrix} \qquad (3)$$

where we have (in the basis $|A_1;v\rangle$ and $|s\rangle$)

$$H(A_1) = \begin{pmatrix} E(A_1;v) & -t(A_1) \\ -t(A_1) & \varepsilon_s \end{pmatrix} \qquad (4)$$

and (in the basis $|T_2;v\rangle$ and $|p\rangle$)

$$H(T_2) = \begin{pmatrix} E(T_2;v) & -t(T_2) \\ -t(T_2) & \varepsilon_p \end{pmatrix} \qquad (5)$$

The vacancy energies $E(A_1;v)$ and $E(T_2;v)$ are obtained from Green's function calculations [9] of ideal vacancy energies, and are the eigenvalues of $H(A_1)$ and $H(T_2)$ in the limit of ε_s and ε_p being infinite [12]. The energies ε_s and ε_p are determined from atomic energy tables; for example, ε_s is 80% of the difference is s-orbital energies of Ga and As for Ga on the As site in GaAs [8,9]. The coupling parameters $t(A_1)$ and $t(T_2)$ are obtained by fitting calculated [9] bulk antisite defect levels.

B. Surface antisite Defects

We next change one of the four neighbors (atom 4) surrounding the antisite into a vacancy. This is accomplished by allowing the antisite only to interact with the a_1 or σ-like molecular orbital, which has no amplitude on atom 4:

$$|a_1\rangle = (\sqrt{3} \, |A_1;v\rangle + |T_2;v\rangle)/2. \qquad (6)$$

The surface Hamiltonian of the antisite interacting with only three neighbors becomes (in a basis $|a_1\rangle$, $|s\rangle$, and $|p\rangle$)

$$H_{surface} = \begin{pmatrix} \varepsilon(a_1) & -t_1 & -t_2 \\ -t_1 & \varepsilon_s & 0 \\ -t_2 & 0 & \varepsilon_p \end{pmatrix} \quad (7)$$

where $\varepsilon(a_1)$ is the self-energy of the remaining three sp^3-hybrid orbitals, and is given by

$$\varepsilon(a_1) = [3\ E(A_1;v) + E(T_2;v)]/4. \quad (8)$$

The s and p orbitals of the antisite interact with the remaining sp^3-hybrids with reduced strengths $t_1 = \sqrt{3}\ t(A_1)/2$ and $t_2 = t(t_2)/2$.

In this simple model, the changes due to the surface are contained in the facts that (a) the s and p orbitals of the antisite interact with an "average" hybrid orbital of its neighbors, having "average" energy $\varepsilon(a_1)$, and (b) the strengths of the interaction for the surface are reduced from those of the bulk. Both effects, in particular (b), can markedly shift the surface antisite levels from those of the bulk.

The results of this simple model are compared with the full surface Green's function calculation in Figs. 1 and 2. For the cation-on-the-anion-site defect (e.g., Ga_{As}), the model yields only one level in or very near the band gap -- an acceptor level that can produce Fermi-level pinning and Schottky barrier formation for an n-type semiconductor. The detailed calculations [6] also produce only a single prominent level in the band gap for this defect -- again an acceptor level. When the results of the

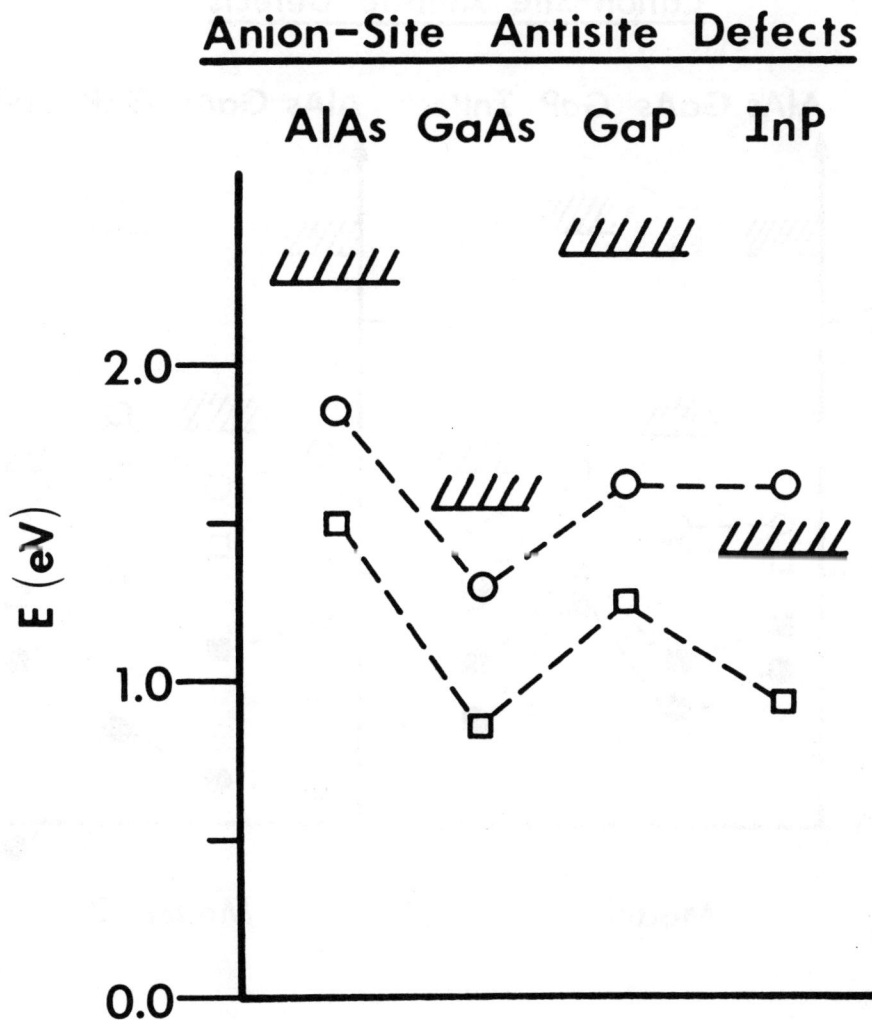

Fig. 1. Predictions of simple four-atom model for acceptor levels associated with cation-on-anion-site antisite defects (open circles) compared with predictions of detailed calculations (open squares) [6].

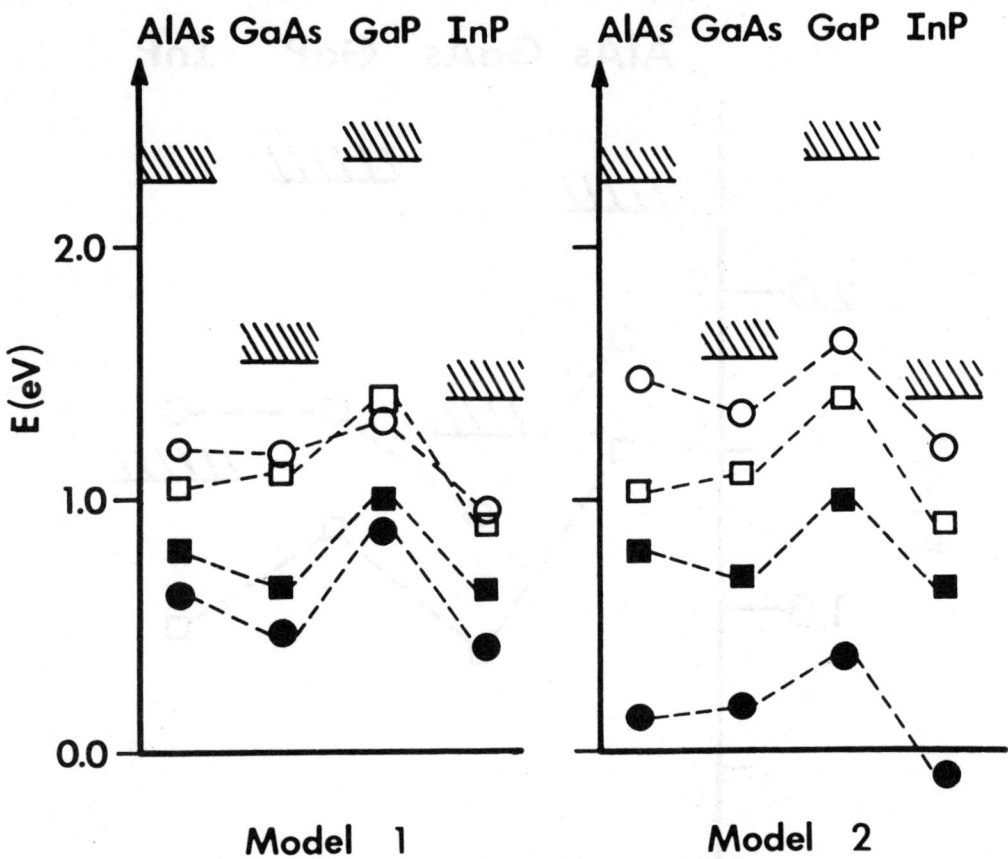

Fig. 2. Predictions of simple four-atom model for acceptor and donor levels associated with anion-on-cation-site antisite defects (circles) compared with predictions of detailed calculations (squares) [6]. Open circles and squares are acceptor levels (empty for neutral defect), and solid circles and squares are donor levels (filled for neutral defect). "Model 2" represents the "exact" calculation for the four-atom model, and "Model 1" represents a calculation in which the indirect coupling between s-orbital and dangling-bond p-orbital on the defect site is neglected. (See text.)

simple model (open circles) and detailed calculations (open squares) are compared in Fig. 1, it can be seen that the chemical trends are in remarkably good agreement.

A similar comparison for the other antisite, the anion-on-the-cation-site defect, is shown in Fig. 2. As described above, two versions of the simple model were used: In "model 1," the 3×3 problem of Eq. (7) was artificially decoupled to yield the two 2×2 problems of Eqs. (4) and (5). This amounts to neglecting the indirect interaction between the defect-site s-orbital and dangling-bond p-orbital via the direct interaction of each of these orbitals with neighboring orbitals on the adjacent anion atoms. (See Eq. (7).) In "model 2," the full 3×3 problem is solved. In both models, one acceptor level and one donor level are produced in (or very near) the band gap. (The "better" model, model 1, gives the "worse" results because of the hybridization of s-orbital and dangling-bond p-orbital on the antisite defect; this is not the relevant point, however.) As can be seen in Fig. 2, either of these versions of the simple 4-atom model (open and solid circles) yields chemical trends almost identical to those of the detailed calculations (open and closed squares).

The agreement between the simple models of both defects and the detailed calculations indicates that both approaches provide a reliable description of the chemical trends. It also indicates that these trends have a simple physical origin, principally involving the dangling-bond p-orbital for the anion-site defect (e.g., Ga_{As}) and both the s-orbital and dangling-bond p-orbital for the cation-site defect (e.g., As_{Ga}).

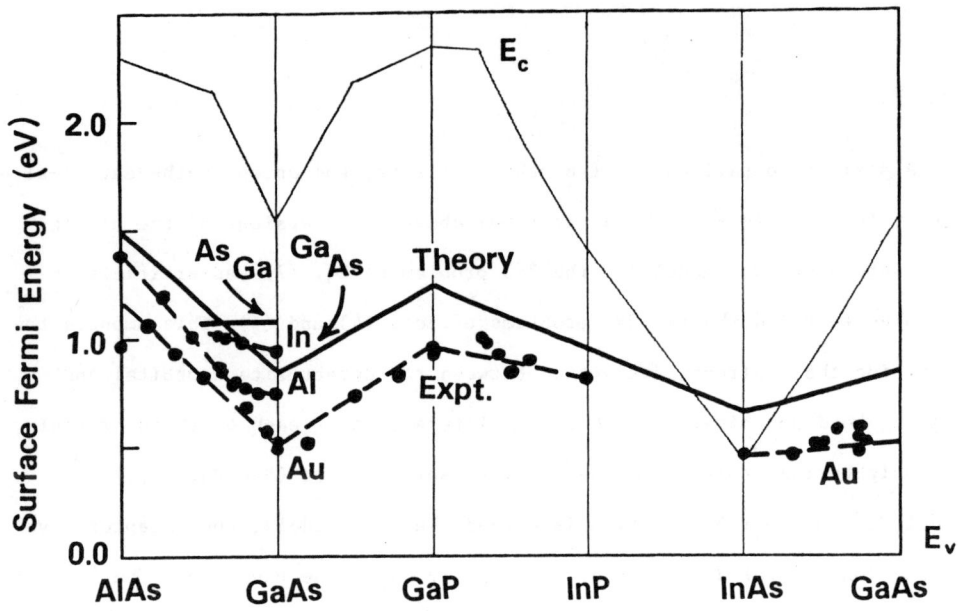

Fig. 3. Predictions of detailed calculations for the cation-on-anion-site defect (e.g., Ga_{As}) and the anion-on-cation-site defect (e.g., As_{Ga}) at relaxed (110) surfaces of III-V semiconductors and their alloys. Only the acceptor levels, relevant to Fermi-level pinnings on n-type semiconductors, are shown. For $Al_xGa_{1-x}As$, note that the slope of the acceptor level (dE/dx) is large for the cation-on-the-anion-site defect and small for the anion-on-the-cation-site defect. The experimental data for Au contacts to various alloys, and Al and In contacts to $Al_xGa_{1-x}As$, are also shown. (References to the experimental papers are given in Refs. [3] and [6]. We attribute the data for Au contacts and for Al contacts to Al-rich $Al_xGa_{1-x}As$, to Fermi-level pinning by the cation-on-the-anion-site antisite defect. The anion-on-cation-site defect is identified as responsible for pinning the Fermi level at In and Al contacts to Ga-rich $Al_xGa_{1-x}As$. These results indicate that the slope of the Fermi-level pinning position as a function of alloy composition (dE/dx) can serve as a signature of the defect type. The conduction band edge is denoted E_c.

In Fig. 3, we show predictions of our detailed calculations for several III-V alloys, compared with experimental Fermi-level pinning positions inferred from measurements of Schottky barriers and MOS (metal-oxygen-semiconductor) structures. (The sources of the experimental data are cited in Ref. [6] and the review of Mönch [3].) For Au contacts to all alloys, the data appear to be well described by the cation-on-anion-site defect level (e.g., Ga_{As}). This defect state is cation dangling-bond-like in character and draws its strength mainly from the conduction band. Hence its energy changes considerably as the alloy composition varies.

The data for In and $A\ell$ for small x in $A\ell_x Ga_{1-x} As$ however show only a modest change with alloy composition. In fact, $A\ell$ appears to produce a kink in the Fermi-level pinning position as a function of x. The defect model readily explains this behavior in terms of a "switching" of the dominant defect from the cation-on-anion-site defect (e.g., Ga_{As}) for large x to the anion-on-cation-site defect (e.g., As_{Ga}) for small x. The anion-on-cation-site defect level has anion dangling-bond character, is valence-band-like, and hence shows little change as the alloy composition x varies.

III. SUMMARY

Thus the simple picture of Fermi-level pinning by deep energy levels associated with defects accounts for the chemical trends in the Schottky barrier data. Indeed, the essential physics is contained in the simple four-atom model which, in a hybrid basis, can be easily evaluated.

We thank the Office of Naval Research (Contract Nos. N00014-84-K-0352 and N00014-82-K-0447) for their generous support.

REFERENCES

[1] J. Bardeen, Phys. Rev. $\underline{71}$, 717 (1947).

[2] W. E. Spicer, P. W. Chye, P. R. Skeath, C. Y. Su, and I. Lindau, J. Vac. Sci. Technol. $\underline{16}$, 1422 (1979); W. E. Spicer, I. Lindau, P. R. Skeath, and C. Y. Su, J. Vac. Sci. Technol. $\underline{17}$, 1019 (1980); W. E. Spicer, I. Lindau, P. R. Skeath, C. Y. Su, and P. W. Chye, Phys. Rev. Letters $\underline{44}$, 520 (1980); and references therein.

[3] W. Mönch, Surf. Sci. $\underline{132}$, 92 (1983) and references therein.

[4] R. H. Williams, Surf. Sci. $\underline{132}$, 122 (1983) and references therein.

[5] H. H. Wieder, Inst. Phys. Conf. Ser. $\underline{50}$, 234 (1980); Appl. Phys. Lett. $\underline{38}$, 170 (1980).

[6] R. E. Allen, T. J. Humphreys, J. D. Dow, and O. F. Sankey, J. Vac. Sci. Technol. B$\underline{2}$, 491 (1984). See also R. E. Allen and J. D. Dow, Phys. Rev. B$\underline{25}$, 1423 (1982).

[7] O. F. Sankey, R. E. Allen, and J. D. Dow, Solid State Commun. $\underline{49}$, 1 (1984).

[8] P. Vogl, H. P. Hjalmarson, and J. D. Dow, J. Phys. Chem. Solids $\underline{44}$, 365 (1983).

[9] H. P. Hjalmarson, P. Vogl, D. J. Wolford, and J. D. Dow, Phys. Rev. Letters $\underline{44}$, 810 (1980).

[10] R. E. Allen, Phys. Rev. B $\underline{20}$, 1454 (1979).

[11] S. Y. Ren, W. M. Hu, O. F. Sankey, and J. D. Dow, Phys. Rev. B$\underline{26}$, 951 (1982).

[12] M. Lannoo and P. Lenglart, J. Phys. C $\underline{30}$, 2409 (1969).

PHASE TRANSITIONS IN THE LAPLACIAN ROUGHENING MODEL

David A. Bruce
Department of Physics and Astronomy
Michigan State University, East Lansing, MI 48824-1116

ABSTRACT

The Laplacian roughening model is studied in one and two dimensions. The simulation in 2D supports the theories of Halperin and Nelson, and of Young, which explain the melting of a 2D solid by a two-step process. The mechanisms of the surface transitions are explained in terms of the surface topology. The surface undergoes a roughening transition caused by the growth of linear steps, and a disorienting transition caused by the formation of a random network of folds.

1. INTRODUCTION

The Laplacian roughening model [1] owes its existence, not to the need to represent the essential qualitative features of a real physical system, but rather to the fact that it may be mapped by a duality transformation onto another model which does arise naturally in the study of two-dimensional solids. In particular, it maps onto a lattice gas model which has been used to describe the phenomenon of "two-dimensional melting" [2]. Although there may be other applications for the Laplacian model, in this paper it will be discussed only with respect to the melting problem.

The statistical mechanics of two-dimensional models has received a great amount of attention since the tour de force of Onsager's exact solution of the Ising model. Soon after this breakthrough, Burton, Cabrera, and Frank (BCF) [3] introduced the concept of solid on solid (SOS) models in studies of crystal surface growth. They suggested that an SOS model would exhibit a phase transition of a novel kind in which at a critical temperature the surface would be roughened [4] by the spontaneous nucleation of steps, crossing the surface. Considerable interest was also shown in two-dimensional lattice gases, principally the Coulomb gas which was found to undergo a metal-insulator transition. It was only in 1975 that Chui and Weeks [5] discovered that the partition function of the Coulomb gas problem could be mapped onto the partition function for a particular SOS model, the discrete Gaussian (DG) model. This showed that the roughening and metal-insulator transitions are in the same universality class. The principal advantage which results from this procedure is that the lattice gas and the SOS models are amenable to different theoretical and numerical simulation techniques, therefore a study of the dual model, or of a model in the same universality class as the dual model, may yield information on a problem that a direct study of the physically relevant model would not. The models in the SOS roughening universality class have been very well understood as a result of extensive Monte Carlo simulations, a technique for which the integer arithmetic arising in the models is ideal, and through the observation by van Beijeren [6] that one of the family, the body-centered, or BCSOS model is a physical realization of the exactly soluble six vertex models [7]. Although the Coulomb lattice gas may be solved by an approximate renormalization group technique [8], this connection to the roughening models has reinforced our understanding of many details of the metal-insulator transition.

The phenomenon of two-dimensional melting is usually explained by a lattice gas model in which the lattice gas charges are either dislocations or disclinations, point-like defects or topological singularities in the crystalline structure. Again, this model may be solved approximately at low particle densities, however recent controversies over the nature of the transition, or transitions, have made it natural to study the models involved through their dual, SOS-like counterpart, the Laplacian model.

In this work, section 2 will contain a brief review of the roughening and melting problems. This will be followed in section 3 by a description of the duality transformation as applied to the Laplacian model. A brief outline of the solution in one dimension will be presented in section 4, and then in the concluding sections the major results gained by computer simulation of the two-dimensional problem will be discussed.

2. The Roughening and Melting Problems
2.1 SOS Models and Roughening

In a solid on solid model, the growth of a three-dimensional crystal is modelled by placing cubic unit cells on a square lattice. The cubes are placed on top of each other so that the surface is completely specified by a set of integer height variables $\{h(\vec{r}_i)\}$ where the $\vec{r}_i$ are the lattice vectors of the two-dimensional square lattice, see Figure 1. Although the original calculations were carried out for models in which the heights $h(\vec{r}_i)$ could range over only two values (an Ising model) or over a small number of layers, $h(\vec{r}_i)$ is usually assumed to vary over all integer values from $-\infty$ to $+\infty$. The surface fluctuations are controlled by a Hamiltonian, $H(\{h(\vec{r}_i)\})$, which contains only short-range forces, typically involving only the height difference between adjacent columns.

Surface roughness is defined by the asymptotic behavior of a correlation function $G(\vec{r}_i - \vec{r}_j)$, where

$$G(\vec{r}_i - \vec{r}_j) = \langle(h(\vec{r}_i) - h(\vec{r}_j))^2\rangle$$

the angle brackets denote the usual canonical ensemble average.

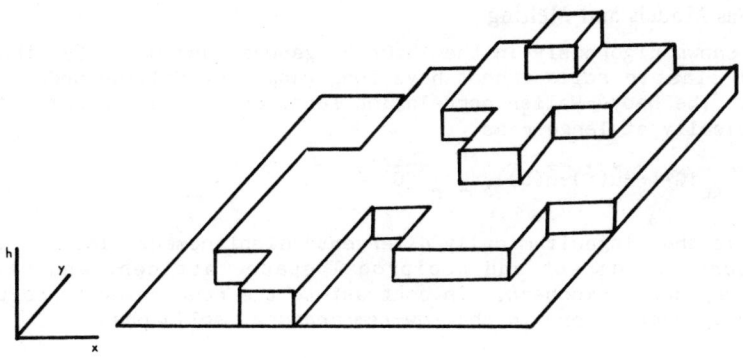

Fig. 1: Sketch of part of the surface of an SOS model. Although the heights are in a vertical direction the lattice is considered two-dimensional in the X-Y plane.

In the limit of large $|\vec{r}_i - \vec{r}_j|$ we have the condition

$$\lim_{|\vec{r}| \to \infty} G(|\vec{r}|) = \text{constant} \Rightarrow \text{surface is smooth}, \; T < T_R$$

$$\lim_{|\vec{r}| \to \infty} G(|\vec{r}|) = \infty \qquad \Rightarrow \text{surface is rough}, \; T \geq T_R$$

where T_R is the roughening temperature. If $h(\vec{r})$ is a continuous variable, then capillary waves cause the surface to be rough at all temperatures. If $h(\vec{r})$ is a discrete variable, then for a large class of "conventional" roughening models

$$G(|\vec{r}|) \sim \ln\left(\frac{|\vec{r}|}{\xi}\right) \; : \; T \geq T_R \; .$$

The transition may be viewed as a breaking of the continuous (vertical) translational symmetry by the discrete translational steps imposed by the integer heights. An alternative correlation function, the generating function $C_\alpha(|\vec{r}|)$ has the suggestive form

$$C_\alpha(|r|) = \langle e^{i\alpha(h(\vec{r}) - h(\vec{0}))} \rangle \sim \gamma^{-\eta(T)} \; : \; T > T_R$$

showing power law decay of correlations in the high-temperature phase with a temperature-dependent exponent $\eta(T)$.

The transition is rather subtle, and approximate theoretical treatments such as mean field theory and Migdal-Kadanoff-type renormalization group are not sensitive enough to pick up the logarithmic divergences. The former suggested that there might be no transition before computer simulations confirmed the original work of BCF. It will be seen later that the weakness of the thermodynamic singularities causes some difficulty in interpreting the Laplacian model results.

2.2 Lattice Gas Models and Melting

It was shown rigorously in the 1960s by general arguments [9] that a two-dimensional elastic solid cannot have long-range crystalline order at finite temperature. The Debye-Waller correlation function for such a solid tends to zero algebraically at large $\vec{r}$ as

$$C_{\vec{G}}(\vec{r}) = \langle e^{i\vec{G}\cdot(\vec{r}+\vec{u}(\vec{r})-\vec{u}(o))} \rangle \sim r^{-\eta_{\vec{G}}(T)}$$

where $\vec{u}(\vec{r})$ is the (logarithmically divergent) displacement field, $\vec{r}$ and $\vec{G}$ are lattice vectors for direct and reciprocal space lattices, and $\eta_{\vec{G}}(T)$ is a temperature-dependent exponent. In contrast to the roughening transition, the algebraic decay occurs here in the low-temperature, solid phase.

Kosterlitz and Thouless (KT) noted that similar behavior, namely low-temperature algebraic decay of correlations occurred in the 2D X-Y model, and that it persisted up to a temperature at which vortices could be excited in the spin system [10]. The KT vortices are point defects which interact logarithmically, and whose energy is proportional to the logarithm of the system size. The KT transition may be found approximately by a simple energy-versus-entropy-balance argument, or more accurately by means of a renormalization group calculation. This theory was extended and applied to the melting of an elastic solid by Halperin and Nelson and by Young (HNY) [2], who noted that the nucleation of logarithmically interacting dislocations would cause a transition, as a temperature T_m, above which the Debye-Waller correlation function would have the expected exponential decay

$$C_{\vec{G}}(|\vec{r}|) \sim \exp(-|\vec{r}|\xi) \quad : \quad T > T_m .$$

HNY showed that this disclination-unbinding transition is not sufficient to cause the solid to melt into an isotropic liquid, as angular or orientational correlations with algebraic decay persist above T_m. Rather, the phase immediately above T_m may be termed a hexatic or tetratic liquid crystal for a partially melted hexagonal or square lattice respectively. In order to melt the hexatic phase, HNY assumed that a second transition involving the creation of disclinations would occur at a higher temperature T_I, above which the solid would have melted into an isotropic liquid and both translational correlations and orientational correlations would decay exponentially at large distances. The transitions at T_m and T_I are both continuous in the sense that a vanishingly small number of free dislocations (disclinations) are present just above $T_m(T_I)$, however elastic constants behave discontinuously, having "universal jumps" at the transitions.

An appealing description of the two phase transitions can be given solely in terms of disclinations, since a dislocation can be viewed as a pair of tightly-bound disclinations. In this picture, the elastic solid at low temperature is expected to contain a low density of tightly-bound disclination quadrupoles, with no free dipoles (dislocations) or isolated disclinations. The pairwise interaction potential for disclinations in an elastic solid varies with the separation of the "charges" n_i and n_j as [11]

$$V(\vec{r}_{ij}) \approx \frac{K}{16\pi} n_i n_j |\vec{r}_{ij}|^2 \ln|\vec{r}_{ij}|$$

where K is an elastic constant, and the disclination charge, or disclinicity n_i is defined to be

$$n_i = \frac{z}{\epsilon\pi} \oint d\theta(\vec{r})$$

where z is the coordination no. of the lattice and $\theta(r)$ the bond angle field. The density of disclination is controlled by their core energy. Separating a quadrupole into two dipoles and then separating the dipoles by a distance r, gives an interaction

$$V_D(\vec{r}_{ij}) \approx \frac{K}{4\pi} \vec{b}_i \cdot \vec{b}_j \ln|\vec{r}_{ij}|$$

where the vectors $\vec{b}_i$ and $\vec{b}_j$ are just the Burgers vectors of the dislocations formed by the disclination pairs. These logarithmically interacting dipoles undergo a KT transition at T_m leading to the hexatic phase. In the hexatic phase, the inter-disclination potential cannot be calculated exactly. It may be inferred from the HNY results that the dipole plasma existing in the hexatic phase screens the interaction potential for inter-disclination interactions from the tightly binding $r^2 \ln r$ dependence, to a $\ln r$ dependence. The now logarithmically interacting disclinations will undergo a KT unbinding transition at the higher transition temperature T_I.

While this description of two-dimensional melting is persuasive, it is not rigorous. The calculation of screening effects close to the phase transitions rests on the approximate KT renormalization group method which is valid only at low dislocation or disclination density. A simulation by Saito [12] of a dipolar lattice gas supposed to represent a gas of dislocations in a melting solid, suggested that a crucial parameter in the theory is the dislocation core energy, E_c. For large core energies, Saito found that the transition at T_m was continuous and in good agreement with HNY theory, however for a lower core energy, he found that the transition became discontinuous, with the abrupt formation of dislocation loops at T_m, presumably indicating that the two-stage continuous melting process had been pre-empted by a single, first order solid-liquid transition.

3. The Dualty Transformation and the Laplacian Model
3.1 Transforming the Partition Functions

The duality transformation of Chui and Weeks was introduced to map the discrete Gaussian model partition function onto that of the Coulomb gas. The transformation can be applied in almost the same way to the Laplacian model.

The Laplacian model is defined as a SOS model with the Hamiltonian

$$H_l(\{h(\vec{r}_i)\}) = \frac{1}{2} J \sum_i (\Delta h(\vec{r}_i))^2 + H \sum_i h(\vec{r})^2.$$

where Δ is a discrete lattice approximation to the Laplacian operator ∇^2, which for the square lattice with which we will be concerned in this paper,

$$\Delta(h(\vec{r}_i)) = \left(\sum_{<nn>} h(\vec{r}_j) \right) - 4h(\vec{r}_i).$$

The notation indicates that $\vec{r}_j$ is a nearest neighbor site to $\vec{r}_i$. The partition function is then the usual configurational sum

$$Z_L = \sum_{\{h(\vec{r}_i)\}} \exp - \beta\, H_L\left(\{h(\vec{r}_i)\}\right)$$

The field H appears in the Hamiltonian only to control trivial divergences, during calculations we will always be interested in the limit H tends to 0. In this limit the Laplacian model has two discrete symmetries. There is an infinite set of zero-energy, ground states where $h(\vec{r}_i)$ is constant, the discrete translational symmetry of the SOS models. There is also an infinite set of ground states where the surface is planar, but oriented at an angle to the horizontal with an integer slope. The presence in the model of these two different discrete symmetries, translational and orientational, together with the assumption that at high temperature the discrete nature of the height variables ought to be unimportant, in which case the surface has neither a well-defined position nor orientation, suggests that there may be two phase transitions in this model, one associated with locking into a particular orientation, and one at a lower temperature associated with locking into a particular position.

To carry out the duality transformation, we replace the sum over $h(\vec{r}_i)$ in the partition function Z_L by an integral over a continuous variable h, weighted by a set of delta functions, i.e.

$$W = \prod_{i=1}^{N} \sum_{\{n(\vec{r}_i)\} = -\infty}^{\infty} \delta(h(\vec{r}_i) - n(\vec{r})).$$

The h variables are now integrated out of the partition function leaving an expression which involves only the integers $n(\vec{r}_i)$. With some rearrangement this becomes

$$Z_L = Z_U \sum_{\{n(\vec{r}_i)\}} \exp\left[-\frac{1}{2} \tilde{J}\, \tilde{\beta} \sum_{i,j} n(\vec{r}_i)\, n(\vec{r}_j)\, U\,(\vec{r}_i - \vec{r}_j)\right]$$

where Z_U is the (non-singular) partition function for an uncoupled lattice gas,

$$\tilde{J} = \frac{4\pi^2}{J}$$

$$\tilde{\beta} = 1/\beta$$

and the lattice-gas-like potential of interaction is

$$U(\vec{r}) = \frac{1}{N} \sum_{\vec{q}} \frac{e^{i\vec{q}\cdot\vec{r}} + \frac{1}{2}(\vec{q}\cdot\vec{r})^2 - 1}{(1 - \frac{1}{2}\cos q_x - \frac{1}{2}\cos q_y)^2 + H}$$

$$\underset{H\to 0}{\approx} \frac{1}{8\pi} r^2 \ln r.$$

The connection between the disclination gas model of 2D melting and the Laplacian roughening model is now apparent. The partition function for the roughening model maps onto that for the disclination gas in an elastic solid with the temperature scales inverted by the transformation. We expect that the dislocation unbinding and disclination unbinding transitions at T_m and T_I will correspond in the Laplacian model to disorienting and roughening transitions at T_2 and T_1.

There is one further aspect of the transformation which is vital in this application. The transformation generates two neutrality constraints on the lattice gas. These are

$$\sum_i n(\vec{r}_i) = 0 \quad : \text{charge neutrality}$$

$$\sum_i \vec{r}_i\, n(\vec{r}_i) = \vec{0} \quad : \text{dipole neutrality.}$$

These two conditions are required if the disclination gas in the elastic solid has been formed by a realistic kinetic process such as the formation of dislocation pairs at a point in the solid.

There is one important limitation inherent in the model as presented, and that is that the core energy of a disclination is fixed to be

$$E_c = \frac{BK}{16\pi}$$

where, denoting Euler's constant by γ,

$$B = \frac{1}{2}\ln 8 + \gamma - 1 = 0.6169.$$

This in turn fixes the dislocation core energy to be

$$E_D = \frac{(B + \frac{1}{2})K}{8\pi}.$$

In the units used by Saito [12] this gives a core energy $E/J = 0.56$, which is very close to the value, $E/J = 0.57$ for which he found a first order transition. The core energy could be varied by adding to the Hamiltonian a term involving higher derivatives of the surface profile, however, as will be explained in section 6, it is doubtful whether this would change the order of the transitions.

3.2 Correlation Functions and Potentials

In the same way that the partition function may be transformed by duality to relate a roughening model to a lattice gas, Chui and Weeks showed that there is a relation between the surface correlation function and the screened lattice gas interaction potential. Since the Laplacian model has the orientational symmetry where the entire surface may be tilted it is necessary always to measure correlation functions which embody this symmetry, therefore rather than use the two-point height-height correlation function of section 2, we consider four-point correlation functions. If $s(\vec{r}_i)$ is a variable equal to ± 1, the Chui and Weeks [5] relation gives

$$\langle (\sum_{i=1}^{4} s(\vec{r}_i) h(\vec{r}_i))^2 \rangle = 2\tilde{\beta}\, W_{MF}(\{s(\vec{r}_i)\}).$$

The left-hand side of this equation is the surface correlation function which may be measured in a simulation, on the right-hand side, $W_{MF}(\{s(\vec{r}_i)\})$ is the effective (screened) interaction potential between a set of infinitesimal test charges of sign $s(\vec{r}_i)$ placed on the sites $\vec{r}_i$ in the lattice gas at temperature $\bar{\beta} = 1/\beta$. In order to satisfy the requirement that the correlation function have the full translational and orientational symmetry of the partition function, the charges $\{s(\vec{r}_i)\}$ must satisfy the charge and dipole neutrality constraints.

In order to study the dislocation and disclination unbinding transitions it is sufficient to look at two correlation functions, $G_S(\vec{r})$ and $G_D(\vec{r})$ given by

$$G_S(\vec{r}) = \langle (h(r,o) - h(o,o) + h(o,r) - h(r,r))^2 \rangle$$

$$G_D(\vec{r}) = \langle (h(\vec{b}) - h(\vec{o}) - h(\vec{r}+\vec{b}) + h(\vec{r}))^2 \rangle$$

$G_S(\vec{r})$ represents four disclination charges on the vertices of a square of side r. It is composed of the potentials of single charges interacting in pairs, or in the roughening model language it measures height-height correlations. Thus it will diverge at a normal roughening transition, equivalent to a disclination unbinding. $G_D(r)$ represents four charges on the vertices of a narrow parallelogram. It is a dipole-dipole correlation function which measures the interaction between dislocations, or in the roughening model language it measures slope-slope correlations. This will not diverge at a normal roughening transition, however, it will diverge at large r if the surface undergoes a disorienting transition. The phase diagrams for the two models and the expected behavior of the correlation functions are shown in the table. The most straightforward method of determining the transition temperatures, and the nature of the screened potentials is simply to simulate the roughening system using a Monte Carlo technique, and calculate the correlation functions equivalent to the appropriate unbinding mono- or dipoles.

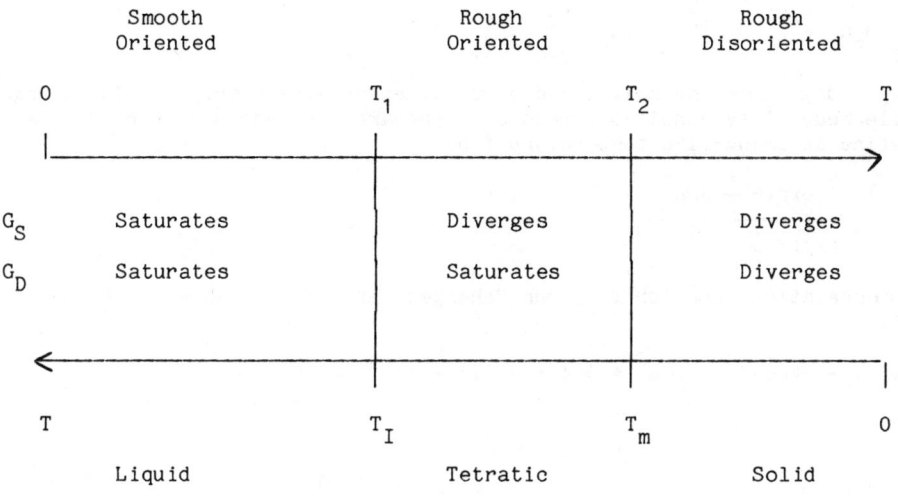

The phase diagrams and inverse temperature scales of the Laplacian roughening model and its dual model, the two-dimensional elastic solid.

4. Exact Solution in one Dimension

It is known that a linear chain with only short-range interactions cannot develop long-range order at any finite temperature. Nevertheless, it is instructive to solve the Laplacian model in one dimension to show the different behavior of the orientational and translational correlation functions as the zero temperature "transition" is approached.

For our linear chain the Hamiltonian is

$$H_L = \frac{1}{2} J \sum_i \Delta_i^2 = \frac{1}{2} J \sum_i (2h_i - h_{i+1} - h_{i-1})^2 .$$

In computing the trace in the partition function it is sufficient to change variables and regard the Δ_i as independent. If periodic boundary conditions were to be applied this would impose a constraint on the Δ_i variables and the trace cannot be carried out. In the thermodynamic limit it is appropriate to use free boundary conditions in which case the partition function factorizes to give simply

$$Z = \left(\sum_{n=-\infty}^{\infty} \exp(-\frac{1}{2} \beta J n^2) \right)^N .$$

Similarly, the four-point correlation function is calculated by writing the operator in terms of the Δ variables in the form

$$\sum_{c=1}^{4} s_c h_c = \sum_c s_c \{h_o - c g_o - \sum_{j=0}^{c-1} \sum_{k=1}^{j} \Delta_k\}$$

where h_o and g_o are the height and slope at an arbitrary origin. The charge and dipole neutrality conditions remove the apparent dependence on h_o and g_o. If we define an Ising-like temperature t by

$$t = \frac{\sum n^2 \exp(-\frac{1}{2} \beta J n^2)}{\sum \exp(-\frac{1}{2} \beta J n^2)}$$

then a correlation function for four "charges" arranged as shown in Figure 2, is simply

$$G(\ell, \delta) = 2t[\frac{1}{3} \delta^3 + \frac{1}{2} \delta^2 + \frac{1}{6} \delta + \frac{1}{2} \ell(\delta + 1)^2] .$$

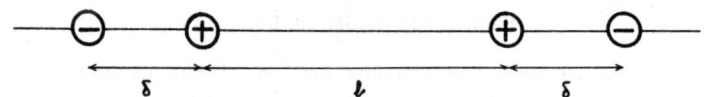

Fig. 2: Arrangement of charges corresponding to 1D correlation function $G(\ell, \delta)$.

The two correlation functions which are relevant to the melting problem are now $G_S(\ell)$, the correlation function for four equally-spaced points representing the "dislocation" potential, and $G_D(\ell)$, the correlation function for two unit dipoles, representing the "disclination" potential. These are given by

$$G_S(\ell) = \frac{5}{3} t \ell^3 \quad ; \quad G_D(\ell) = t \ell .$$

In one dimension these correlation functions are always strongly divergent at large ℓ, the corresponding lattice gas charges are therefore always tightly-bound quadrupoles. Although there is no phase transition, we do have distinct critical scaling behavior for the two types of fluctuation close to t = 0, the translational correlation length scales as $t^{-1/3}$, while the orientational correlation length scales as t^{-1}.

5. Monte Carlo Simulation of the Two-Dimensional Model
5.1 Correlation Functions

The SOS models are ideally suited to Monte Carlo simulation by the Metropolis method. Most of the arithmetic required involves only integer variables and short-range interactions which minimize run time and storage requirements. The calculations described here were carried out on a VAX minicomputer.

From the calculated correlation functions, it is clear even from runs on samples of only 32 x 32 lattice spacings that there are two phase transitions and that these are continuous. The results shown in the accompanying diagrams were actually obtained for a 64 x 64 lattice, the largest lattice for which relaxation times were short enough to allow thermal equilibrium to be maintained with reasonable equilibration times of around 10^5 MC steps per site.

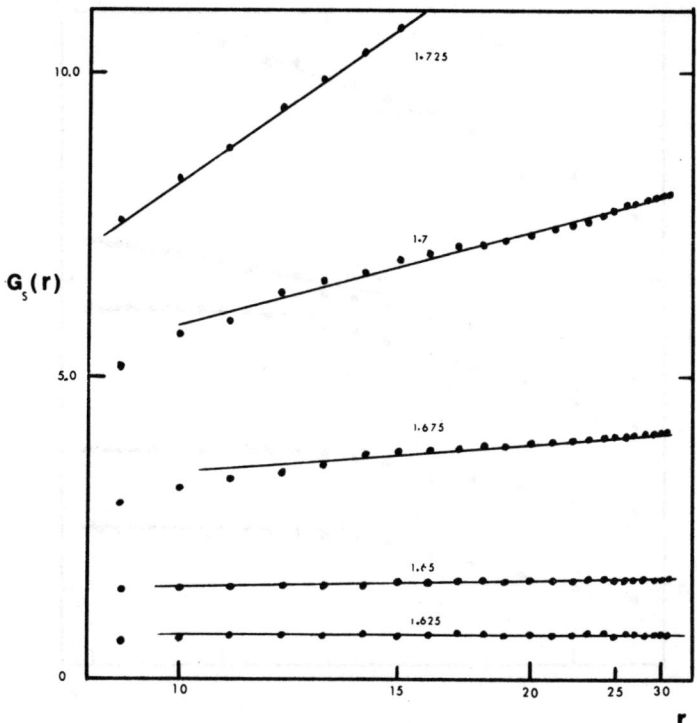

Fig. 3: $G_s(\vec{r})$ for a 64 x 64 lattice at temperatures, close to T_1, shown beside the curves. The x ordinate is the lattice Greens function, the finite system analog of the expected $\ln r$ behavior, to cancel finite size effects, see [16]. The straight lines are a guide to the eye.

The disclination unbinding transition occurs at a temperature T_1 = 1.64 J. Above this temperature the inter-disclination potential can be seen from Figure 3 to be accurately logarithmic for distances greater than about 10 lattice spacings. This confirms the HNY result that the tight binding $r^2 \ln r$ inter-disclination potential is screened by the dislocations to a weak binding $\ln r$ potential in the tetratic phase. It also confirms that the transition at T_1 is an ordinary, and hence well-understood, roughening transition. Thus, the higher temperature transition in the melting model, the transition for which HNY theory is clearly more suspect, has been shown to belong to a universality class for which extremely accurate simulations, and exact solutions are available, such is the power of duality transformations.

The higher temperature transition is more difficult to study by this method. The transition temperature, T_2, is approximately 2.2 J. The difficulty in locating T_2 accurately comes from the need to obtain good statistics for the correlation function in the range greater than 15 lattice spacings, see Figure 4. Close to T_2 we find very long relaxation times, largely due to the amount of phase space which must be sampled to take account of the highly-textured nature of the disordered surface.

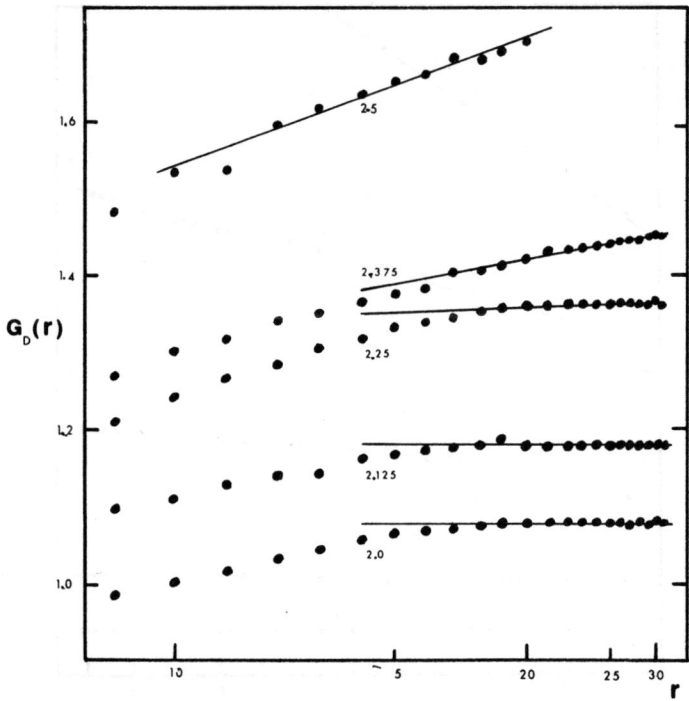

Fig. 4: $G_D(\vec{r})$ for a 64 x 64 lattice at temperatures, close to T_2, shown beside the curves. As in Fig. 2 the x ordinate is the lattice Greens function.

5.2 Monte Carlo Renormalization Group

Since the two transitions, the roughening transition at T_1 and the disorienting transition at T_2, are continuous transitions with divergent correlation lengths, we may apply a renormalization group approach [13]. We calculate a set of effective Hamiltonians by coarse-graining the lattice, effectively carrying out a repeated Kadanoff block-spin transformation, and then look for fixed points of the transformation. By duality, we expect from the approximate KT renormalization group calculations of HNY, that the model should show fixed lines rather than fixed points. The KT fixed lines occur at low temperatures in the dual model, implying that there should be fixed lines in the Laplacian model for temperatures above the phase transition temperatures.

The blocking rule adopted was the simplest possible consistent with retaining all of the symmetries of the model. The four height variables on the corners of a unit square were averaged to give a block height variable

$$h'(\vec{r}) = \frac{1}{4} \sum_{c=1}^{4} h(\vec{r}_i) .$$

$h'(\vec{r})$ must be an integer variable to retain the discrete symmetries, h' values were always rounded to the nearest integer, in the case of ties the rounding up or down was chosen randomly.

It is convenient to look at the effect of the transformation on the energy of the system. Since the model has just a one-parameter Hamiltonian, the energy may be thought of as the effective temperature of the coarse-grained surface. If $T^{(m)}$ is the temperature after m iterations then generally if $dT^{(m)}/dm < 0$, the system is ordered and will flow to a $T = 0$ fixed point, if $dT^{(m)}/dm > 0$, the system is disordered, and if $dT^{(m)}/dm = 0$, the system is at a fixed point. It can be shown that the energy of the Laplacian model should have the following behavior if predicted fixed lines of the HNY approximate renormalization group exist. In the low temperature smooth, oriented phase $E^{(m)}$ should flow to zero. The tetratic, or rough but oriented phase, will be marked by a fixed energy $E^{(m)}$ which is independent of m. The high temperature rough and disoriented phase will have $E^{(m)} \alpha\, b^{2m}$ where b is the rescaling length parameter, in this case 2. This behavior is shown schematically in Figure 5.

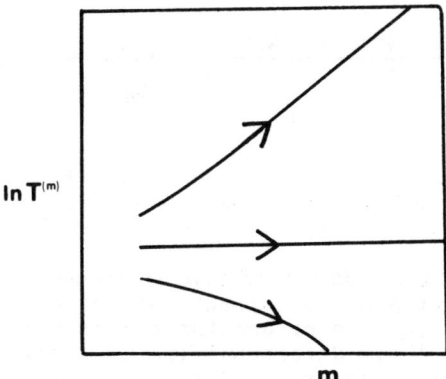

Fig. 5: Sketch of the predicted RG flow of the effective temperature $T^{(m)}$.

The results obtained for simulations on a 32 x 32 lattice are shown in Figure 6. The low temperature phase is clearly marked by the flow toward zero at large m of the energy. The intermediate phase is marked by an energy which increases monotonically with m rather than remaining constant. Since the rate of increase is decreasing, this may be due to a transient which will ultimately be renormalized away at large m, however, within the size of sample which it is possible to work with, it is evident that the expected asymptotic behavior is not reached. The upper transition is not detectable by the MCRG method. The RG flows at very high temperature show the anticipated behavior, and it seems that in principle T_2 is marked by a change in the curves in Figure 6 from concave downward to concave upward, however, close to T_2 the data is too inaccurate to place the transition with any confidence.

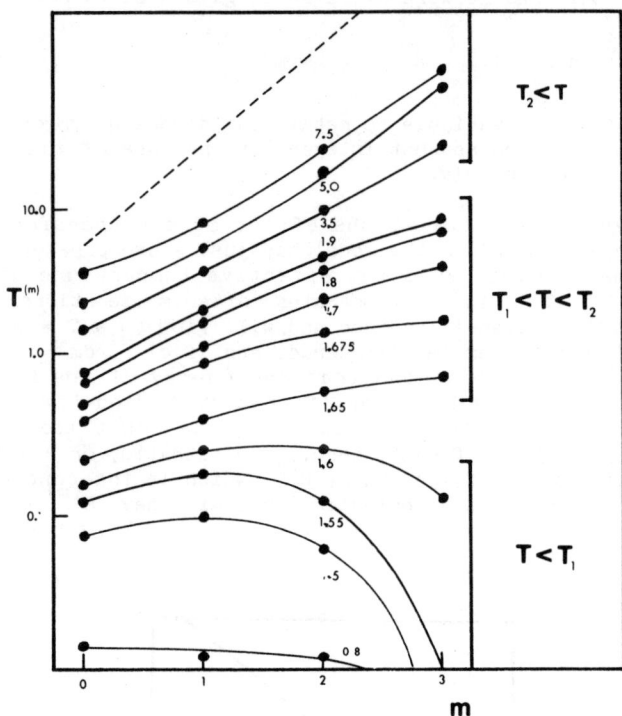

Fig. 6: $T^{(m)}$ versus m for a 32 x 32 lattice at the temperatures shown beside the curves. The dashed line shows the slope in the expected high temperature limit.

The MCRG calculation gives an alternative procedure for fixing T_1 which agrees well with that obtained by direct inspection of the correlation functions. It also shows that there are three phases marked by different asymptotic behavior of $E^{(m)}$. Unfortunately, the large transients observed even below T_1 show that the Laplacian model's behavior is dominated by surface features, i.e. steps, facets, or raised islands whose size is comparable to the size of the systems in the simulation. This contrasts with situation in the discrete Gaussian model where the MCRG flows show no detectable transients.

6. The Mechanisms of the Phase Transitions

It is at first sight surprising that the simple one parameter Laplacian Hamiltonian should give rise to two transitions, quite widely spaced in temperature. The transitions can be quite well understood qualitatively by looking at the actual surface features in the phases above T_1.

In the rough but oriented phase close to T_2 the surface profile shows large flat regions bounded by steps. The transition mechanism is clearly the vanishing free energy of a step at T_1 due to a balance between its energy and entropy as originally conjectured by BCF. If we compare the Laplacian and DG models, the energy per unit length of a straight step in the Laplacian model is J. With this energy per unit length, the roughening temperature for the DG model is $T_n \approx 1.48$ J. In the Laplacian model, unlike the DG model, there is an additional energy cost of J for each kink in a step, so that the Laplacian model has a tendency to form slightly more regular, larger structures than the DG model and so we would expect the roughening temperature to be raised slightly. The value $T_1 = 1.64$ J is quite consistent with this picture.

In the rough, disoriented phase above T_2 the surface profile shows sloping planar facets, small pieces of the various low-order ground states, bounded by fold lines. To form a fold along a (1,0) direction costs an energy of only 1/2 J per site; however, to create folds along (1,1) directions costs 2 J per site, and folds in other directions which would cost even higher energies do not appear. The network of folds join up at vertices where two (1,0) type folds and one (1,1) type fold meet. The resultant pattern is reminiscent of the "topological froths" studied [14] in connection with soap films. Since it is necessary, in order to disorient the surface, to create the folds with energy 2 J compared with the steps which cost only J which are required to roughen the surface, it is to be expected that the disorienting temperature T_2 would be higher than the roughening temperature T_1.

The two continuous transitions obtained with an effective core energy for which Saito's [12] results predict a single, first-order transition is clearly in disagreement with the dislocation gas simulation. The role of the lattice gas core energy in the transitions of the dual roughening model is not clear, the transition temperatures here being determined by the energy and entropy of linear steps or random fold networks. It is possible in this model to change the effective core energy by adding a higher derivative term to the Hamiltonian H_L. The effect of this would be to change the energies per site of a step or fold, and hence to change the transition temperatures T_1 and T_2. It is not possible, however, by this means to lower the disorienting transition temperature, since lowering the energy of a low order fold such as the important (1,1) folds necessarily causes a greater reduction in energy of the higher order folds so that very high order folds will be energetically favored even at zero temperature, destroying the phase transitions altogether. On the square lattice, the disorienting transition will thus always be at a higher temperature than the roughening transition. In a similar Monte Carlo study of the triangular lattice Laplacian roughening model [15], which has a slightly different effective core energy and a different surface fold topology the hexatic phase was found to be much narrower in temperature than in this study of the square lattice model. The difference probably arises from the fact that only low energy (1,0) type folds are required to disorient the surface in this case. Even in this case adding a term to alter core energies will not produce a single first-order transition.

7. SUMMARY AND CONCLUSIONS

By simulating the Laplacian roughening model, we have obtained directly the screened inter-disclination potential for the dual disclination lattice gas. By inspection of these potentials and by other means we have shown that the models undergo two phase transitions, confirming the predictions of the HNY theory. The mechanisms by which the transitions take place may be thought of as the formation of steps or a network of folds on the surface at T_1 and T_2 respectively. These mechanisms do not appear to be consistent with the predicted first-order transition at low core energy.

ACKNOWLEDGEMENTS

The author would like to thank D.R. Nelson who suggested this problem, Harvard University for its hospitality where most of this work was carried out, the SERC and NSF for their support, and the many people with whom I enjoyed useful discussions in the course of the work.

REFERENCES

1) Nelson, D. R.: Phys. Rev., B, 1982, <u>26</u>, 269.

2) Halperin, B. I. and Nelson, D. R.: Phys. Rev. Lett., 1978, <u>41</u>, 121; Nelson, D. R. and Halperin, B. I.: Phys. Rev. B, 1979, <u>19</u>, 2457; Young, A. P.: Phys. Rev. B, 1979, <u>19</u>, 1855.

3) Burton, W. K. and Cabrera, N.: Disc. Faraday. Soc., 1949, <u>3</u>, 33; Burton, W. K., Cabrera, N. and Frank, F. C.: Phil. Trans. Roy. Soc., 1951, <u>243A</u>, 299.

4) For a review see Weeks, J. D. in "Ordering in Strongly Fluctuating Condensed Matter Systems" ed. T. Riste (Plenum, New York) 1979, p. 293.

5) Chui, S. T. and Weeks, J. D.: Phys. Rev. B, 1976, <u>14</u>, 4978.

6) van Beijeren, H.: Phys. Rev. Lett., 1977, <u>38</u>, 993.

7) Lieb, E. H.: Phys. Rev. Lett., 1967, <u>18</u>, 1046; Sutherland, B.: Phys. Lett. A, 1968, <u>26</u>, 532.

8) Kosterlitz, J. M.: J. Phys. C, 1974, <u>7</u>, 1046 and 1977, <u>10</u>, 3753.

9) Peierls, R. E.: Ann. Inst. Henri Poincare, 1935, <u>5</u>, 177; Mermin, N.D.: Phys. Rev. 1968, <u>176</u>, 250.

10) Kosterlitz, J. M. and Thouless, D. J.: J. Phys. C 1973, <u>6</u>, 1181.

11) Nabarro, F.R.N. Theory of Dislocations (Clarendon, New York) 1967.

12) Saito, Y.: Phys. Res. B, 1982, <u>26</u>, 6239: the first order transition was predicted by Chui S. T.: Phys. Rev. Lett. 1982, <u>48</u>, 933.

13) The method used here largely follows Tobochnik, J.: Phys. Rev. B, 1982, <u>26</u>, 6201.

14) Weaire, D. and Rivier, N.: Contemp. Phys., 1984, <u>25</u>, 59.

15) Strandburg, K. J., Solla, S. A. and Chester, G. V.: Phys. Rev. B, 1983, <u>28</u>, 2717.

16) Saito, Y. and Müller-Krumbhaar, H.: Phys. Rev. B, 1981, <u>23</u>, 308.

INTERFACIAL PHASE DIAGRAMS AND EQUILIBRIUM CRYSTAL SHAPES

Michael Wortis
Department of Physics and Materials Research Laboratory
University of Illinois, Urbana, IL 61801

ABSTRACT

Equilibrium crystal shapes are a manifestation of interfacial thermodynamics. The evolution with temperature of the location of edges on the equilibrium crystal defines an interfacial phase diagram. Such phase diagrams are calculated for Kossel crystals with nearest- and next-nearest-neighbor interactions. Universality classes of the second-order phase boundaries are identified and compared with recent experimental data.

INTRODUCTION

When a macroscopic single crystal is left for sufficiently long at firstorder coexistence with its melt or vapor at temperature T, it attains an equilibrium crystal shape (ECS). The ECS consists in general of some combination of strictly planar facets and smoothly curved surfaces, meeting at edges which may be sharp (slope discontinuity) or smooth (no slope discontinuity). It has been known since the classic work of Wulff [1], Herring [2], and others that the ECS reflects the fact that the free-energy cost $f_i(\hat{m},T)$ of making a unit area of solid/fluid interface at temperature T depends on the interface orientation $\hat{m}$ relative to bulk-crystal axes. The geometrical construction which connects $f_i(\hat{m},T)$ to the function $r(\hat{n},T)$ which describes the ECS by specifying the radius r from the center of the crystal to its surface in the direction $\hat{n}$ is called the "Wulff construction."

Interfacial Phase Diagrams

It was recently pointed out by Andreev [3] that the Wulff construction is simply a two-dimensional version of the usual Legendre transform used in thermodynamics [4] to pass from one thermodynamic potential to another. Thus, $r(\hat{n},T)$ is an interfacial thermodynamic potential and the relation $\hat{m}(\hat{n},T)$ which specifies the tangent plane to the ECS in direction $\hat{n}$ is an equation of state. It follows that the locus of singularities of the function $r(\hat{n},T)$ constitutes an interfacial phase diagram [5]. These singularities are just the edges occurring on the ECS. Sharp edges (discontinuity in first derivatives of r) correspond to first-order phase transitions; smooth edges (no discontinuity in first derivatives) correspond to second-order phase transitions. While all values of $\hat{h}$ appear on the ECS, in the presence of first-order edges certain orientations $\hat{m}$ are absent. Thus, there may be forbidden regions in the ($\hat{m}$,T) phase diagram (locus of singularities of f_i) but not the ($\hat{h}$,T) phase diagram, i.e. $\hat{m}$ is a "density variable," while $\hat{h}$ is a "field variable."

The thermal evolution of the ECS of some simple cubic Kossel crystals (lattice crystals with nearest-neighbor and next-nearest-neighbor interactions) has recently been worked out via mean-field [5,6] and solid-on-solid-model [7,8,9] techniques. The result for nearest-neighbor interactions (only) is shown as Fig. 1 below, along with the corresponding ($\hat{h}$,T) phase diagram (restricted to the equatorial plane). At T=0 the ECS is a cube. As T is increased, (100) facets remain but are now separated by smoothly curved surfaces. The facet/curved-surface edges are second-order. When T reaches the (100) roughening temperature T_R, the facets shrink to zero, and above T_R the ECS is a single smoothly rounded surface without edges. Inclusion of further-neighbor interactions [5,9] produces, depending on sign, either additional facets or first-order phase transitions.

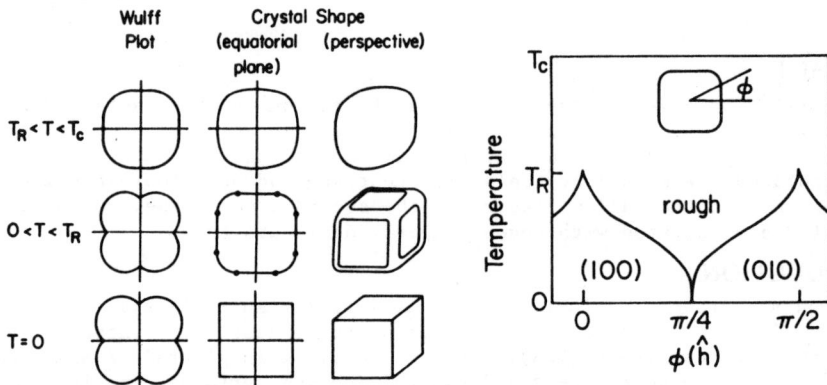

Figure 1. Equilibrium shape of nearest-neighbor Kossel crystal as a function of temperature T. Evolution with T of edge position in the equatorial plane is shown in the phase diagram at the right. T_R and T_c are the roughening and critical temperatures, respectively.

Interfacial Phase Transitions

The interfacial phase transitions associated with facet-edge shapes for models of the Kossel-crystal type (with short-range forces) can be related to the known behavior of certain two-dimensional statistical mechanical models [5-9]. In particular:
(a) The second-order edges separating facets and curved surfaces (which appear in Fig. 1 for $0 < T < T_R$) are associated with a Pokrovsky-Talapov transition [10]. This means that tangency is predicted to be of the form $y \sim x^\theta$ with $\theta=3/2$ (instead of the quadratic tangency $\theta=2$ suggested by

mean-field theory [3]). This critical exponent appears to have been observed for micron-scale lead crystals [11].

(b) Behavior near T_R is of the Kostlitz-Thouless type [12]. There is predicted [7,8,9] to be a jump in interface curvature in the (100) direction from a finite universal value for $T=T_R^+$ to zero for $T=T_R^-$. Such a jump, analogous to the jump in the d=2 superfluid density [13], has been recently observed for ^{4}He crystals near 1 K [15]. In addition, the facet area should vanish as $T \to T_R^-$ like $\exp(-c/\sqrt{T_R-T})$, a behavior which is constant with observations on Ag_2S [16].

REFERENCES

1) Wulff, G.: Z. Kristallogr. Mineral. (1901) 34, 449.
2) Herring, C.: Phys. Rev. (1951) 82, 87.
3) Andreev, A. F.: Zh. Eksp. Teor. Fiz. (1981) 53, 2024 [Sov. Phys. JETP (1982) 53, 1063].
4) Callen, H. B.: Thermodynamics (Wiley, New York, 1960), p. 93.
5) Rottman, C. and Wortis, M.: Physics Reports (1984) 103, 59.
6) Rottman, C. and Wortis, M.: Phys. Rev. B (1984) 29, 328.
7) Jayaprakash, C., Saam, W. F., and Teitel, S.: Phys. Rev. Lett. (1983) 50, 2017.
8) Saam, W. F., Jayaprakash, C., and Teitel, S.: in Quantum Fluids and Solids-1983, edited by E. D. Adams and G. G. Ihas, AIP Conference Proceedings No. 103 (American Institute of Physics, New York, 1983), p. 371.
9) Jayaprakash, C. and Saam, W. F.: "Thermal Evolution of Crystal Shapes: The FCC Crystal," Ohio State University preprint (1984).
10) Pokrovsky, V. L. and Talapov, A. L.: Phys. Rev. Lett. (1979) 42, 65.
11) Rottman, C., Wortis, M., Heyraud, J. C., and Métois, J. J.: Phys. Rev. Lett. (1984) 52, 1009.
12) Kosterlitz, J. M. and Thouless, D. J.: J. Phys. C (1973) 6, 1181; José, J., Kadanoff, L. P., Kirkpatrick, S., and Nelson, D. R.: Phys. Rev. B (1977) 16, 1217.
13) Nelson, D. R. and Kosterlitz, J. M.: Phys. Rev. Lett. (1978) 40, 1727.
14) Gallet, F., Wolf, P. E., and Balibar, S.: "Universal Jump of the Curvature of ^{4}He Crystals at the Roughening Transition," Ecole Normale Supérieur preprint, 1984.
15) Ohachi, T. and Taniguchi, I.: J. Cryst. Growth (1983) 65, 84.

This work was supported in part by the National Science Foundation under Grants No. DMR81-17182 and DMR83-16981.

THE SELECTION PRINCIPLE OF DENDRITIC SOLIDIFICATION
THE SNOWFLAKE PROBLEM

Eshel Ben-Jacob
Institute for Theoretical Physics
University of California
Santa Barbara, CA 93106

and

Department of Physics
The University of Michigan
Ann Arbor, MI 48109

ABSTRACT

We review the phenomenological model of dendritic solidification (the boundary layer model or BLM) proposed by Ben-Jacob, Goldenfeld, Langer and Schön. The BLM incorporates interfacial kinetics, crystalline anisotropy and a local approximation for the dynamics of the thermal diffusion field. We show that the dynamically selected velocity and tip radius of the dendrites in the boundary-layer model have the special values which permit the existence of steady state needle-crystal solutions. This result provides new insight concerning the role of crystalline anisotropy and the validity of the marginal-stability hypothesis.

1. INTRODUCTION
1.1 Motivation

Everywhere around us are nonlinear dissipative systems far from equilibrium. However, we still have a long way to go before we can understand the fundamental principles underlying the dynamics and selection rules of processes in such systems. Seeking to gain understanding of such problems, we have put a lot of effort into modelling the complex pattern formation in snowflakes [1] (Fig. 1) and unraveling the deeper link between this and other processes of selection and pattern formation. Pattern formation occurs in a wide variety of nonlinear dissipative systems driven beyond the limits of stability of their spatially homogeneous states. Recently attention has focussed on the mechanisms of wavelength selection [2-5] in a variety of systems exhibiting periodic spatial structures, including hydrodynamic instabilities in fluids [6], electrohydrodynamic instabilities of nematic liquid crystals [7], cellular flame fronts [8], autocatalytic chemical reactions [9] and directional solidification [10]. These systems possess families of linearly stationary states of different structure, yet in practice a unique state is reproducibly

Fig. 1   A "typical" real snowflake [10].

selected. Much research in this area has been aimed at finding some stability principle which would select a naturally preferred state from the family of allowed states of the system.

Here we are concerned with dendritic solidification where a localized perturbation (solid) of an unstable state (the undercooled liquid) grows at a unique velocity. A selection criterion which appears to be consistent with observations is that the system somehow chooses the state of marginal stability [10]. This has not been given a firm basis in theory, partially because of the lack of models which are sufficiently complex to retain the essential features, yet simple enough to be mathematically and computationally tractable. The full free boundary problem (stephan problem), even for the simplest case of solidification of a pure substance, has proved to be extremely difficult for either analytic or direct numerical investigation. A new approach to interfacial pattern formation have been introduced recently by two groups: the geometrical model by Brower et al.[11] and the boundary-layer model by Ben-Jacob et al.[12] Both are local models and promise to be applicable not just in the particular example of dendritic solidification, but also in other situations in Physics, Chemistry and Biology.

1.2 Dendritic Solidification

The salient features of dendritic solidification are as follows: We consider a liquid at temperature T_∞ below the equilibrium melting temperature T_M of its solid in contact with a growing solidification front. In the local equilibrium approximation, the surface temperature T_S of a flat interface is T_M, but because of the work performed against surface tension, a curved interface has temperature lower by an amount proportional to the local curvature, namely;

$$T_S = T_M - \left(\frac{L}{C}\right) d_0 \kappa \qquad (1.1)$$

where C is the liquid heat capacity, L the latent heat of the transition liquid-solid, d_0 is a capillary length proportional to surface tension [10] (of the order of 10-20Å) and κ is the curvature of the interface defined in two-dimensional system by

$$\kappa = \frac{\partial \theta}{\partial s} \qquad (1.2)$$

the angle θ and the arc length s are defined in Fig. 2. In three dimensions κ is twice the mean curvature. A crude nonequilibrium condition is that the normal velocity is proportional to the driving force of solidification--the temperature difference. Namely

$$T_S = T_M - \left(\frac{L}{C}\right) d_0 \kappa - \beta(\theta) v_n \qquad (1.3)$$

where v_n is the velocity of the interface along its outward normal (Fig. 1.2) and $\beta(\theta)$ is a function of the angle θ that reflects the crystalline anisotropy.

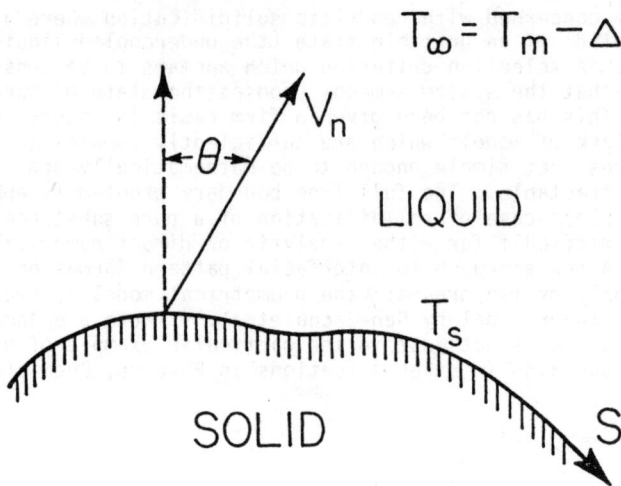

Fig. 2 Schematic illustration of a solidification front, showing various quantities defined in text.

We consider here only dendrites which are formed under rather special growth conditions. We are interested in the case that the only physical mechanism governing the motion of the interface is the diffusion of latent heat. Thus, the undercooling must be sufficiently large to ensure fluid connection effects negligible. Secondly, the undercooling must not be so large that the nucleation and molecular attachment kinetics are the dominant mechanism.

Newly solidifying material at the interface generates latent heat L, which is conducted away from the interface by thermal diffusion (we neglect fluid flow). This process is facilitated by a large interfacial area in contact with the liquid and a large thermal gradient at the interface, both of which may be achieved by the interface developing protrusions extending into the liquid. The surface tension acts as a stabilizing mechanism preventing the formation of deformations on an arbitrarily small length scale, by providing a coupling between the temperature of the interface and the curvature of the deformation. As first shown by Mullins and Sekerka [13], this unstable competition between the thermal diffusion and the surface tension is the underlying mechanism for dendritic growth. The other ingredient essential for the beautiful, regular shapes observed in dendrites is the crystalline anisotropy. Apart from providing the (e.g.) six-fold symmetry seen in snowflakes it may function either as a mechanism channeling the growth of instabilities along crystallographically preferred directions, or as a triggering mechanism for instabilities. Without it, it is likely that the growing dendrite would follow a crooked path rather than propagating along a single direction, giving rise to structures similar to those observed in diffusion-limited aggregation [14]. The effects of crystalline anisotropy arise in part through the kinetics of molecular attachment, an intrinsically non-equilibrium process; its inclusion in a model of dendritic solidification marks a departure from the usual assumption of local equilibrium.

The model described above including the complete dynamics of the thermal diffusion field, defines a highly nontrivial free-boundary problem (FBP) which has so far resisted even direct numerical simulation except in relatively simple special cases. The alternative approach presented below is to derive a boundary-layer model (BLM) for the dynamics of the interface rather than that of the diffusion field, but to take into account in a simple way the basic physical features described above. The boundary layer model is a two-dimensional model of dendritic solidification in which the characteristic decay length of the diffusion field is much smaller than the local radius of curvature of the interface. In this approximation, valid when the dimensionless undercooling is close to unity, thermal diffusion occurs in a thin boundary layer at the interface; variations in the boundary layer thickness account for diffusion perpendicular to the interface.

1.3 The Marginal Stability Hypothesis

The marginal stability hypothesis, proposed by Langer and Muller-Krumbhaar [10], is an attempt to explain the experimental data on the tip velocity and tip radius of dendrites. In the absence of surface tension there is a one-parameter family of steady-state, uniformly translating parabolic solution to the FBP, discovered by Ivantsov in 1947; the velocity and tip radius are not determined uniquely by the undercooling, but are related by ρv = const. When surface tension is taken into account, the naive expectation has been that this relation is preserved approximately, with corrections of $O(d_0)$. Moreover, the inclusion of surface tension has introduced a new length scale in the problem which, in the limit that the scale of the diffusion field is much larger than the tip radius, provides a relation of the form ρv^2 = const. The numerical value of the constant in the latter relation has been estimated for a variety of different geometries--circular, planar, or spherical, paraboloid and paraboliad surface of revolution--based on the idea that the system relaxes to the state which is marginally stable to infinitesimal perturbations. To be precise, the steady state solutions are parameterized by $v\rho^2$; the stability spectrum is then computed for each solution and the eigenvalues ordered according to their real part. Then the state whose largest eigenvalue has real part equal to zero is hypothesized to be the selected state.

This marginal stability hypothesis gives good agreement with experiment, and has been reviewed by Glicksman. It has never been given a foundation in theory, however, and as we will show, cannot be true in the form stated above. We will mention in Section 3.2 how the marginal stability hypothesis may conceivably be a reasonable approximation in some circumstances.

2.1 The Model

The essence of our model is the boundary-layer approximation. That is, we visualize diffusion as occurring within a boundary layer at the solidification front, which is thin compared to the local radius of curvature. For the simple case of a crystal growing in two dimensions, this replaces the full diffusion problem by one-dimensional diffusion along the interface. Diffusion normal to interface is accounted for by variations of the thickness of the boundary layer, ℓ, which satisfies its own dynamical equation. It is more convenient and intuitive to consider, instead of ℓ, the heat content per unit length of the interface.

$$H \equiv T_s \cdot \ell \qquad (2.1)$$

The equation of motion for H, following a point on the interface as it moves along its outward normal, is a statement of heat balance in the boundary layer:

$$\left.\frac{\partial H}{\partial t}\right)_n = v_n \cdot L - C \cdot T_s \qquad (2.2)$$
$$+ D\frac{\partial}{\partial s}\left[\ell \frac{\partial T_s}{\partial s}\right] - \kappa\, v_n H$$

The first term is the latent heat entering the boundary layer, the second term being the fraction not retained by the new layer of solid formed from the liquid. The third term is just the lateral diffusion along the boundary layer where D is the thermal diffusion coefficient. The final term is of purely geometric origin and arises from the change in the differential arclength as the interface grows.

The phenomenological Eq. (2.2) is supplemented by the exact kinematic equations of an interface with normal velocity v_n

$$\left.\frac{\partial \kappa}{\partial t}\right)_n = -\left(\kappa^2 + \frac{\partial^2}{\partial s^2}\right) v_n \qquad (2.3)$$

$$\left.\frac{\partial s}{\partial t}\right)_n = \int_0^s \kappa\, v_n\, ds' \qquad (2.4)$$

with neglect of thermal diffusion in the solid, the rate at which liquid solidifies is the heat current entering the boundary layer from the interface. Thus

$$v_n = -D\frac{\partial T_s}{\partial n} \cong \frac{DT_s}{\ell} \qquad (2.5)$$

equations 2.1 - 2.5 together with the boundary condition eq. (1.3) completely specify the BLM.

Following references 10, 12 we define the dimensionless undercooling Δ by

$$\Delta \equiv (T_M - T_\infty)/(L/c) \qquad (2.6)$$

and the dimensionless temperature

$$U_s = (T_s - T_\infty)/(L/c) \ .$$

It is also natural to measure arclength in units of d_0 and time in units of d_0^2/d. Thus we obtain the following dimensionless equations

$$\frac{\partial h}{\partial t}\Big)_n = v_n(1 - U_s) - v_n \kappa h + \frac{\partial}{\partial s} \frac{h}{U_s} \frac{\partial U_s}{\partial s} \qquad (a)$$

$$(2.7)$$

$$v_r = U_s^2/h \qquad (b)$$

$$U_s = \Delta - \kappa - \beta(\theta)v_n \qquad (c)$$

$$\beta(\theta) = \alpha(1 - \cos(m\theta)) \qquad (d)$$

$$\frac{\partial \kappa}{\partial t}\Big)_n = -(\kappa^2 + \frac{\partial^2}{\partial s^2})v_n \qquad (e)$$

$$\frac{\partial s}{\partial t}\Big)_n = \int_0^s \kappa v_n \, ds' \qquad (f)$$

In our numerical simulations we have used equations 2.7 with reflection symmetry about the tip at $\theta = 0$, so that the first derivatives of h, κ, U_s, v_n vanish there. In the tail of the dendrite, we have also imposed the conditions that these first derivatives vanish. We used a sufficiently long interface that side branches from the tip had not yet propagated to the boundary in the tail. For studying the steady state properties it is convenient to present the derivatives at constant arc length s, Eq. 2.7, than are given by

$$\frac{\partial h}{\partial t}\Big)_s = v_n(1 - U_s) - \kappa_n h + \frac{\partial}{\partial s} \frac{h}{U_s} \frac{\partial U_s}{\partial s} - \frac{\partial h}{\partial s} \int_0^s v_n \kappa \, ds' \qquad (a)$$

$$\left(\frac{\partial k}{\partial t}\right)_s = -\left(k^2 + \frac{\partial^2}{\partial s^2}\right)v_n - \frac{\partial k}{\partial s}\int_o^s v_n k\, ds' \quad \text{(b)}$$

(2.8)

$$v_n = U_s^2/h \quad \text{(c)}$$

$$U_s = \Delta - \kappa - \beta(\theta)v_n \quad \text{(d)}$$

The BLM has been studied analytically and numerically (see Ref. 12 and next section). At a given undercooling, crystalline symmetry and anisotropy strength a unique dendrite-like structure is formed for a large class of initial conditions (after the decay of initial transients). The dendrite consists of a smooth, almost parabolic tip which extends for about five tip radii before forming sidebranches (Fig. 3). The tip propagates without noticeable change of shape, while the sidebranches appear to be generated periodically and to grow out at fixed positions in the laboratory frame, in a manner consistent with experimental observation.

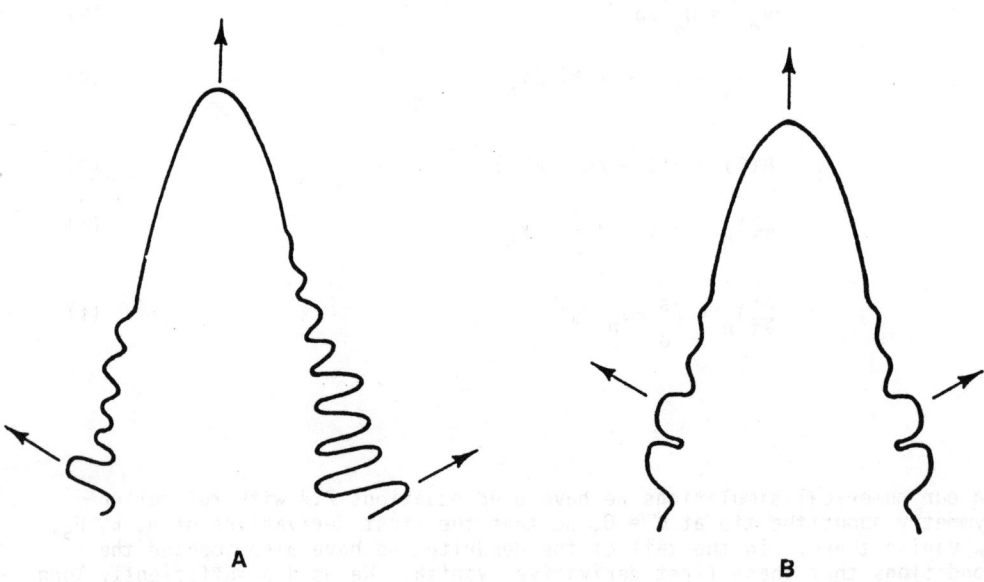

Fig. 3   Qualitative comparison of a real dendriteA(trace of picture of suscinonitrite dendrite from M.E. Glicksman) and B a dendrite produced by the BLM ($\Delta = .8$; $\alpha = .1$; $m = 6$).

DENDRITIC SOLIDIFICATION

At the present we cannot exclude the possibility that the sidebranches seen in the BLM are transients, just as in the GM. We will return to this point in Section 3.2.

2.2 Planar, Circular and Needle Crystal Interfaces

We have been able to verify analytically that the BLM accurately reproduces special solutions of the conventional free-boundary problem and that for $\Delta \sim 1$ the boundary layer is indeed thin compared to the local radius of curvature.

We begin a study of special cases by considering a planar solidification front. Let the displacement of this front be $z(t)$, then

$$\frac{dz}{dt} = v_n = \frac{\Delta^2}{h}$$
$$\frac{dh}{dt} = v_n(1 - \Delta) \qquad (2.10)$$

As long as $\Delta < 1$ the system will settle down after an initial transient to the square-root law characteristic of diffusion controlled growth:

$$z(t) \cong [\frac{2\Delta^8}{1-\Delta} t]^{1/2} \qquad (2.11)$$

which is in agreement with the result of the one dimensional diffusion problem

$$\frac{\partial u}{\partial t} = D \frac{\partial^2 u}{\partial u^2}$$

subject to the continuity condition

$$\frac{\partial u}{\partial z}) = -\frac{v_n}{D} .$$

A very similar analysis can be carried for the two-dimensional case of growing circle. For details see Ref. 12. As shown in this reference an essential feature of the BLM is that it recaptures, quite accurately, the Mullins-Sekerka pattern-forming mechanism in these systems. The stability spectra of planar interface within the BLM are actually quite similar throughout the interesting range of wave numbers to the one obtained from the full diffusion problem.

We turn now to the so-called "needle-crystal" problem, which has generally been considered to be an essential first step in the study of dendritic growth. We shall start by showing how the boundary-layer model recovers almost exactly the well-known family of isothermal solutions first discovered by Ivantsov [15]. The most interesting point to emerge from the following analysis is the fact that, in the boundary-layer approximation, the capillary correction to the Ivantsov model is a singular perturbation which destroys the needle-crystal solutions. Our problem is to find shape preserving growth forms, specifically, for the two-dimensional case, using Eq. (28b), we look for solutions such that

$$v_n = v_0 \cos \theta \qquad (2.12)$$

and

$$\left.\frac{\partial h}{\partial t}\right)_s = 0$$

The Ivantsov limit is obtained by assuming the interface is an isotherm, that is by setting $u_s = \Delta$. Translating this prescription into our Eq. 28, we find

$$\kappa = v_0 \Delta^2 (1-\Delta) \cos^3\theta \qquad (2.13)$$

which is parabola with its tip pointing in the $\theta = 0$ direction.

Parabolic needle crystals in the Ivantsov limit seem to be about as far as one can go in finding exact analytic solutions to the full solidification problem; and it is encouraging that these solutions are recovered almost completely in the boundary-layer model. These solutions are completely unstable, however, as has been shown rigorously both for the full problem and for the boundary-layer model. Thus it is absolutely essential that

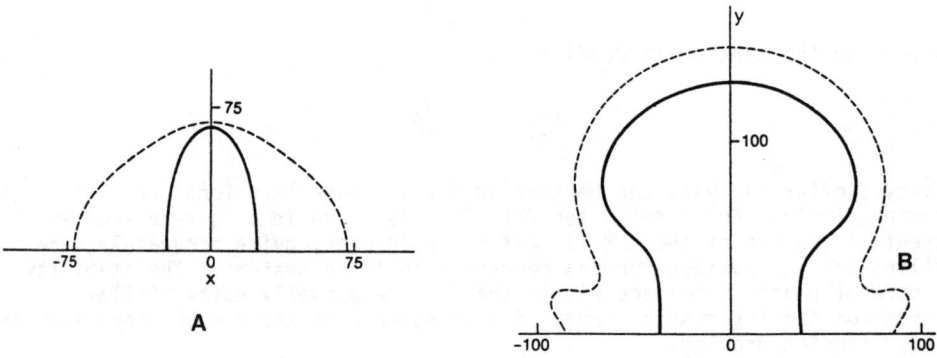

Fig. 4 Evolution of an ellipse at t=0 (A) until t=100 (B) with Δ = .5 and no anisotropy. The solid line is the interface and the dashed line is the edge of the boundary layer.

exact steady-state needle-crystal solutions in this model do not survive the inclusion of capillary corrections without the nonequilibrium term Eq. (1.3). For simplicity we demonstrate this numerically in Fig. 4, for details see Ref. 12.

2.3 Numerical Results

We turn now to a few comments about numerical techniques for dealing with the boundary-layer model.

At first sight, the equations of motion 2.1 - 2.5 appear to constitute a complete system of partial differential equations in one space and one time dimension. Unfortunately this is misleading because the spatial variable s itself obeys an equation of motion, which is coupled to the remaining equations. The time evolution of the discretized equations drives the grid points away from regions of large curvature. This tendency is undesirable not only because the computational error is unevenly distributed along the boundary, but also because of "numerical surface tension". The grid size artificially restricts the length scales probed by the dynamics, and even if there exists a natural short-wavelength cutoff in the system, the modification of the dispersion relation induced by the grid can couple to the long-wavelength behavior via the nonlinearity. In the present example this can lead to an overstabilization of the dendrite tip.

We have dealt with these difficulties by using interpolation to redistribute the grid points evenly along the boundary after each time step. Furthermore, we continuously add new grid points as the boundary expands so that the grid spacing remains constant. The most stable way that we have found is to interpolate the boundary in the space of (s,κ) rather than real space. Our time-step algorithm is a fully implicit scheme with linearization, so that the only limitation on the time step is the rate of change of the quantities we compute. We have verified numerically that the interface undergoes the Mullins-Sekerka instability, the effect being most conspicuous in (s,κ) space, as shown in Fig. 5. In real space, Fig. 6, the interface develops thermal grooves which resemble those seen in undercooled crystals. Ultimately the opposite sides of the grooves intersect, but this is not important from our point of view. Correlations are absent between points close to one another in two-dimensional space, but distant along the boundary, so that the model is strictly accurate only when the groove width is much larger than the width of the boundary layer at the groove walls. It is also interesting to compare the structures generated by the anisotropic boundary-layer model with a real snowflake. We started from a circle and imposed a sixfold anisotropy, with $\Delta = 0.8$. In Fig. 7 we present the time development showing that many of the qualitative morphological features of a real snowflake are indeed reproduced by the anisotropic boundary-layer model.

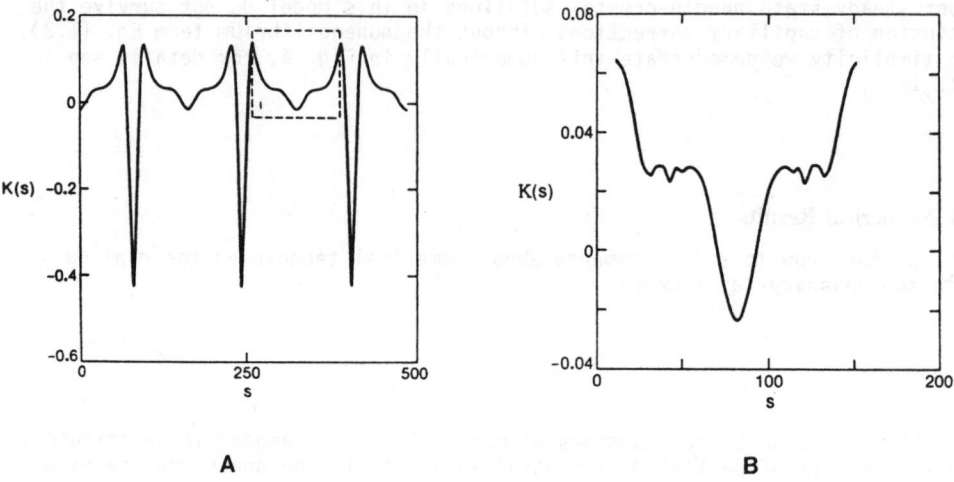

Fig. 5 (a) κ vs. s at t=435 (see Fig. 6) in the small Δ limit, showing the Mullins-Sekerka instability. The structure corresponds to the formation of grooves in real space. (b) Region in (a) enclosed by the dashed square at t=520, showing the emergence of a complex fractal-like pattern in (s,κ) space.

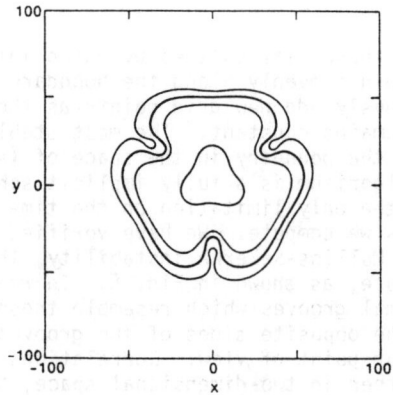

Fig. 6 Time-ordered sequence of the pattern of Fig. 5 plotted in real space, showing grooving. The thick line corresponds to the (s,κ) graph in Fig. 5.

3. The Solvability Selection
3.1 The Phase-Space Analysis

In the absence of surface tension there is a one-parameter family of steady needle crystals--shape-preserving uniformly translating structures resembling dendrites without sidebranches. In this case, the interface is an isotherm, and is completely unstable. When surface tension is included, the traditional assumption has been that a family of steady state solutions persists, but that members of the family with tip radius smaller than a critical length are stable in the moving frame of the tip. The uniqueness and reproducibility of experimental and numerical dendritic structures poses the problems of pattern selection--which of these states is selected and secondly, by what mechanism does the selection occur?

In Ref. 12 we showed that the continuous family of steady-state shape-preserving needle crystals which was assumed to exist on the basis of approximate analyses of the full solidification problem, does not exist in the BLM except in the manifestly unstable Ivantsov limit of vanishing surface tension. Rather, for any set of growth parameters, the BLM has only one or at most a discrete set of needle crystal solutions, each associated with its own growth velocity v_0 and tip curvatures κ_0. Dendrites, with their complex time dependent sidebranching structures, are far from being needle crystals, therefore it might seem unlikely that the dynamical selection mechanism is closely tied to the existence of special stationary solutions which remain needle-like indefinitely far behind the tip [15]. The principle result to be reported here is that the tip of the dynamically selected dendritic mode in the BLM turns out to have precisely the same tip velocity v_0 and tip curvature k_0 as the needle crystal. The existence of the underlying needle crystal solution is determined by matching solvability condition. This is why we call the selection principle a solvability selection. We would like to emphasize that Kessler, Koplik and Levine [17] show that pattern selection in the geometrical model is also tied to the existence of a special needle crystal solution. This makes it plausible that this new selection principle has a wide range of applicability.

First, we look for steady-state shape-preserving solutions by considering the steady-state version of Eq. 2.8 $(\frac{\partial h}{\partial t})_s = 0; (\frac{\partial \kappa}{\partial t})_s = 0)$, which can be written in the form:

$$\frac{d\theta}{ds} \equiv \theta = \kappa = \Delta u_s - \beta v_0 \cos \theta \qquad (3.1)$$

$$\frac{d\omega}{ds} \equiv W = \lambda \qquad (3.2)$$

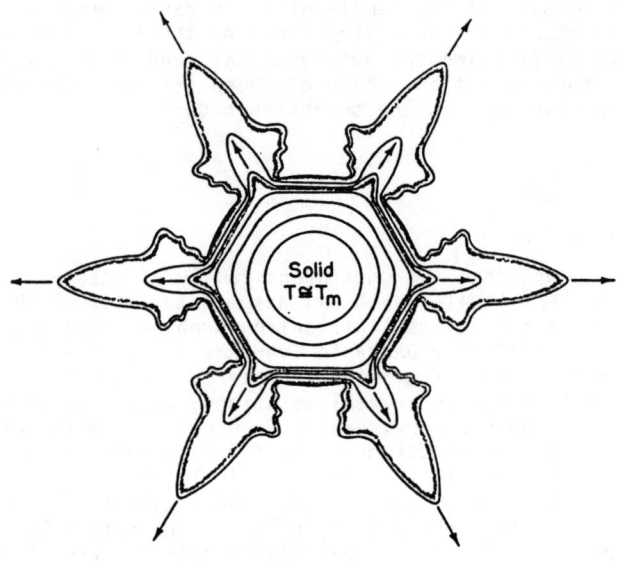

Fig. 7 Time-ordered sequence of the emergence of snowflake-like pattern starting from initial circular seed. The parameters are $\Delta = .8$, $\alpha = .1$, and $m = 6$.

$$\frac{dt}{ds} \equiv \Lambda = 2v_0 \sin\theta\lambda + \frac{v_0 \kappa\omega}{\cos\theta} \left(\frac{1-\Delta\omega}{\omega_s}\right) \quad (3.3)$$

$$- \frac{\lambda^2}{\omega} - \lambda\kappa\tan\theta$$

Here we have used the fact that 2.8b admits a first integral which satisfies the boundary conditions, namely that $v_n = v_0 \cos\theta$ where v_0 is the velocity of the tip.

In order to analyze 3.1 - 3.3 we consider the phase space of θ, w, λ. The only important fixed points are at

$$\theta = \pm \pi/2$$
$$w = 1 \quad (3.4)$$
$$\lambda = 0$$

We emphasize that the tip is not a fixed point ($\frac{d\lambda}{ds} = \frac{d^2w}{ds^2} \neq 0$, since the tip is the point of coldest temperature). To understand the physical meaning of the two fixed points, we have to recall that a steady-state needle-crystal solution of the BLM corresponds to a trajectory in the phase-space which joins the two fixed points and passes through the line $\theta = \lambda = 0$. In physical terms the fixed points correspond to a flat interface at $s \to \pm \infty$ which is an isotherm (λ 0) at the melting temperature ($w = 1$). Linearization about the fixed points (3.4) of Eqs. 3.1 - 3.3 shows that these fixed points are singular. There is only a single trajectory which enters or leaves the fixed points. This trajectory is asymptotically identical to the Ivantsov solution;

$$\kappa(\theta) \underset{\theta \to \pm \pi/2}{\to} v_0 \Delta^2 (1-\Delta) \cos^3\theta \quad (3.5)$$

However, the Ivantsov needle-crystal cannot connect the two fixed points and satisfy the symmetry at the tip since it is an isotherm.

In order to compute the needle crystals solution connection the two fixed points explicitly, we have integrated Eqs. 3.1 - 3.3 backwards in s from $\theta \cong \pi/2$, using Eq. 3.5 to identify initial values of θ, w, λ as close as is numerically feasible to that fixed point. For general values of v_0 the trajectory reaches $\theta = 0$ at some $\lambda \neq 0$. For only discrete values of v_0 the trajectory reaches $\theta = 0$ with $\lambda = 0$ which corresponds to a needle-crystal solution. In Fig. 8 we show $\lambda(\theta = 0)$ as function of ν_0 for $\alpha = .1$ m = 4 and $\Delta = .75$, we have found two eigenvalues of ν_0 at $\nu_0 = .2712$ with $\kappa_0 = .0915$ which has exactly the shape of the dynamically selected dendrite tip (Fig. 9), and another at $\nu_0 = .0215$ with $\kappa_0 = .00545$ which is slow, flat tipped solution.

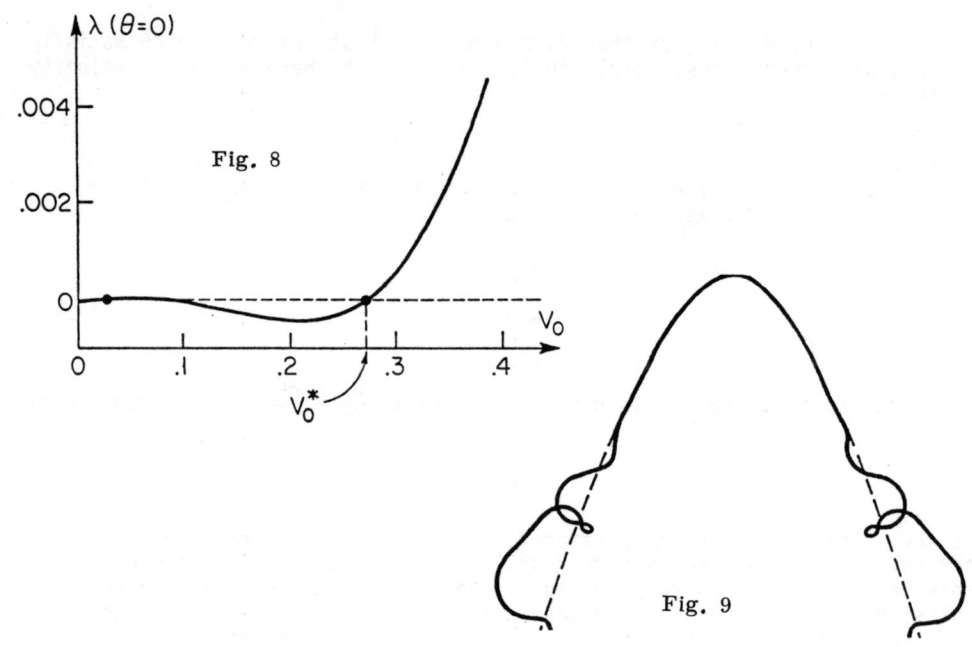

Fig. 8 $\lambda(\theta=0)$ as function of ν_0 for $\alpha = .1$ m = 4 and $\Delta = .75$.

Fig. 9   Comparison of the time-dependent dendrite (thick line) and the underlying "shooting" needle crystal solution (dashed line).

3.2 Interpretation and Application to Stability Considerations

To understand the significance of this result, note first that the needle crystal, which we shall denote by $\kappa^*(s)$, is a functional fixed point of the dynamical system, Eq. (2.8). We shall assume that the fully time dependent solution of (1) is approaching a functional limit cycle, to be denoted by $\kappa^*(s,t)$. The shapes of both theoretical and experimental dendrites seem to repeat themselves with a well defined period when observed in a frame of reference moving with the tip. We cannot prove that our computed $\kappa(s,t)$ in the BLM will not diverge or otherwise lose periodicity at some large but finite time, nor can we discount the possibility that late-stage coarsening of sidebranches occurs (in real dendrites of the BLM) via a weak instability of the motion. But the limit cycle appears to be a good working hypothesis for the moment.

Our numerical results tell us that the limit cycle $\kappa^*(s,t)$ coexists in the dynamical function space with the fixed point $\kappa^*(s)$ and; moreover, that $\kappa^*(s,t)$ and $\kappa^*(s)$ are close to each other in the sense that they are numerically indistinguishable near the tip of the dendrite. We suggest that the simplest way to view this situation is as if $\kappa^*(s,t)$ has emerged through a single Hopf bifurcation. $\kappa^*(s,t)$ and $\kappa^*(s)$ both lie on an invariant two-dimensional manifold with $\kappa^*(s)$ unstable only against perturbations in that manifold. Alternatives to this picture seem improbable. If $\kappa^*(s)$ were completely stable, it would be likely to appear as an attractor in the dynamical simulations; but we do not seem to see needle crystals without sidebranches. If $\kappa^*(s)$ were completely stable, it would be likely to appear as an attractor in the dynamical simulations; but we do not seem to see needle crystals without sidebranches. If $\kappa^*(s)$ were unstable against more than one complex-conjugate pair of deformation modes, then points initially near $\kappa^*(s)$ would not flow reliably toward a unique limit cycle. In neither case could we understand how the properties of $\kappa^*(s,t)$ are accurately determined by the needle crystal $\kappa^*(s)$. Note that, if $\kappa^*(s)$ must have exactly one conjugate pair of unstable modes, then we already have arrived at a weak form of a marginal stability principle. If the modes are part of a continuous spectrum, then $\kappa^*(s)$ decays algebraically rather than exponentially. Alternatively, the modes can belong to a discrete part of the spectrum. Both of the latter possibilities imply that the selected state of dendritic growth is one in which the tip is characterized by a specially weak instability. The first possibility is consistent with the sharp statement that the needle crystal associated with dendritic motion is marginally unstable. We conjecture that this is the case.

As a first step towards testing the above picture, we have linearized Eq. (2.8) about the stationary solution $\kappa^*(s)$ and have studied the eigenvalue spectrum of the resulting operator both analytically and numerically. Details of these investigations will be presented elsewhere, but the crucial result can be understood from qualitative considerations. The needle crystal has the special property that the thermal field h diverges like $s^{1/2}$ as $\theta \to \pm\pi/2, s \to \pm\infty$. In the BLM, this means that all deformation modes which propagate down the dendrite (as observed in the moving frame of the tip) can neither grow nor decay in a linear theory. For example, in the planar stability spectrum reported in Eq. (4.6) of Ref. 12, vanishes for all q in the limit of large h and vanishing normal velocity. It turns out that the extended states that constitute the continuous part of the dendritic deformation spectrum have the asymptotic form $\exp[iqs+\omega(q)t]$ as $s \to \pm\infty$ with $\omega(q) = iqv_0^*$. Thus, all of these modes become stationary in the laboratory frame when observed at positions far from the tip. A similar phenomenon was observed by Müller-Krumbhaar and Langer in their numerical studies of linear stability in the full diffusion problem; the spectrum seemed to flatten out at $\text{Re}\omega \approx 0$ near $q \approx 0$. The slowing of all modes rather than just those of long wavelength as $h \to \infty$ is an exaggerated feature of BLM. Our conclusion is that, in both the BLM and the full diffusion problem, any exact needle crystal solution can be no more stable than marginally stable, and the mode (or modes) with $\text{Re}\omega = 0$ must lie in a continuous part of the stability spectrum.

The remaining portion of the spectrum must consist of modes which are localized near the tip of the dendrite. Our numerical simulations of the dynamical system, Eq. (2.8) plus preliminary results of a numerical stability analysis lead us to believe that these modes either merge with the continuum at $\text{Re}\omega = 0$ or become stable with $\text{Re}\omega < 0$ for sufficiently large anisotropy strength α. Because of the intrinsic marginal stability of the needle crystal, we conjecture that dendritic behavior occurs in both the BLM and the full diffusion problem throughout some nonzero range of values of α.

We conclude with some remarks concerning the status of the marginal stability hypothesis. The statement of a marginal stability criterion suggested by our new results is different from previous statements in an important way: the stability requirement pertains to the needle-crystal fixed point and not to the dendritic limit cycle. This is a weakening of the hypothesis in the sense that it no longer says anything about dynamical behavior not associated with a fixed point. In another sense, however, the statement is surprisingly strong because marginal stability is simultaneously a criterion for the existence of a dentritic limit cycle and a property of the needle crystal. In practical terms, this means that if one has a family of approximate needle-crystal solutions, then it is reasonable to select one member of this family by testing for marginal stability, which is exactly what has been done previously. On the other hand, the natural procedure would be to find exact needle crystals and use stability as a consistency requirement.

ACKNOWLEDGEMENTS

This research was done in collaboration with Nigel Goldenfeld, B.G. Kotliar, J.S. Langer and G. Schön. We were benefitted from useful discussions with A. Karma and G. Dee.

This research was supported by U.S. Department of Energy Grant No. DE-FG03-84ER54108 and by the National Science Foundation Grant No. PHY77-27084 supplemented by funds from the National Aeronautics and Space Administration.

REFERENCES

[1] J. Kepler "DeNive Sexangula" Godrey Tampach, Frankfort (1611).

[2] L. Kramer, E. Ben-Jacob, H. Brand, M.C. Cross, Phys. Rev. Lett. $\underline{49}$, 1891 (1982).

[3] G. Ahlers, D. Cannel, Phys. Rev. Lett. $\underline{50}$, 1583 (1983).

[4] A.C. Newell, M.C. Cross, preprint.

[5] E. Ben-Jacob, to be published.

[6] J.P. Gollub in "Nonlinear Dynamics and Turbulance" eds. D.D. Joseph and G. Looss, Plenum Press (1983).

[7] R.W.H. Kozlogski, E.F. Carr, Mol. Cryst. Liq. Cryst. $\underline{64}$, 299 (1981).

[8] G.I. Sivashinsky, Ann. Rev. Fluid Mech. $\underline{15}$, 179 (1983).

[9] P.C. Fife, "Mathematical Aspects of Reacting and Diffusing Systems", Springer N.Y. (1979).

[10] J.S. Langer, H. Muller-Krumbhaar, Acta Metall. $\underline{26}$, 1681, 1689, 1697, (1978) J.S. Langer, Rev. Mod. Phys. $\underline{52}$, 1 (1980).

[11] R. Brower, D. Kessler, J. Koplik, H. Levine, Phys. Rev. Lett. $\underline{51}$, 1111 (1983).

[12] E. Ben-Jacob, N. Goldenfeld, J.S. Langer, G. Schön, Phys. Rev. Lett. $\underline{51}$, 1930 (1983).

[13] W.W. Mullins, R.F. Sekerka, J. Appl. Phys. $\underline{34}$, 323 (1963) and $\underline{35}$, 444 (1964).

[14] T.A. Witten, Jr., L.M. Sander, Phys. Rev. Lett. $\underline{47}$, 1400 (1981).

[15] G.P. Ivantsov, Dokl. Av.ad. Nauk. SSSR $\underline{58}$, 567 (1947). See also G. Horvay, J.W. Cahn, Acta Metall. $\underline{9}$, $\overline{695}$ (1961).

[16] E. Ben-Jacob, N. Goldenfeld, B.G. Kotliar, J.S. Langer, preprint.

[17] D. Kessler, J. Koplik, H. Levine, preprint.

CRYSTALS DEFECTS IN CURVED THREE-DIMENSIONAL SPACE

Joseph P. Straley
Department of Physics and Astronomy
University of Kentucky, Lexington, KY 40506

ABSTRACT

The surface of a four-dimensional sphere is a curved three-dimensional space in which it is possible to have a close packing of spheres with perfect icosahedral symmetry. This paper presents some results for defect structures in the icosahedral crystal.

Topological Frustration

There is both a short range and a long range aspect to the stabilization of a crystalline structure. The local structure should give minimum energy; but the example of close packing of disks on the surface of a large sphere shows that this is not sufficient: the local geometry would favor the hexagonal geometry of the planar packing, but since the surface is not flat, strains build up which inevitably must disrupt the order[1]. Nelson[2] has coined the phrase "topological frustration" to describe this incompatability between local and large scale order.

Surprisingly, the packing of spheres in three dimensions can be regarded as another example: on the basis of local geometry one would expect to find packings based on the icosahedron and the tetrahedron; but long range structures built solely of these elements are impossible, because the tetrahedron dihedral angle is not a submultiple of 360°, and the edge of a icosahedron is 5% greater than its circumradius. The direct consequences are that there are two close packing geometries in three dimensions, neither of which is as symmetric as the icosahedron. Another consequence may be the very existence of glassy phases in three dimensions[3].

Changing the curvature of a space changes the frustration. Thus is it interesting to consider the possible arrangements of particles confined to the surface of a four dimensional sphere of radius R: a set of points (w_i, x_i, y_i, z_i) such that $w_i^2 + x_i^2 + y_i^2 + z_i^2 = R^2$, which includes Euclidean space as the limit R --> ∞. Decreasing R decreases the frustration until R = (1+√5)/2, where there occurs a structure of 120 particles such that each particle has twelve neighbors at unit spacing, arranged with icosahedral symmetry. In what follows this "icosahedral crystal" (which is more properly called the {3,3,5} polytope [4]) will be regarded to be the perfect unfrustrated crystal of which the usual three-dimensional crystal structures, confined to the uncomfortable geometry of Euclidean space, are badly distorted images.

In another publication[3] the idea that topological frustration is relevant to the existence of glasses was investigated by Monte Carlo annealing of random arrangements of particles with the potential

$$V(r) = 4/r^{12} \qquad (1)$$

(where the factor of 4 facilitates comparison with studies of the Lennard Jones potentials[5]). It was found that 120 particles on the surface of the hypersphere would form a crystal upon annealing, whereas this did not happen for 108 particles in a 3 x 3 x 3 cubic box with periodic boundary conditions. The purpose of this paper is to explain the geometry of the {3,3,5} polytope and related defect structures.

The Icosahedral Crystal and other Structures

One possible representation of the hypersphere surface is given by the coordinate system (θ,λ,ϕ) such that

$$\begin{aligned} w_i &= \cos\theta\cos\lambda \\ x_i &= \sin\theta\cos\lambda \\ y_i &= \cos\phi\sin\lambda \\ z_i &= \sin\phi\sin\lambda \end{aligned} \qquad (2)$$

The regular polytope of 120 particles is given by the choices[4]

$\lambda = 0$	$\theta = (2\mu+1)\pi/10$	ϕ = arbitrary	(3a)
$\lambda = \arctan 1/2$	$\theta = 2\mu\pi/10$	$\phi = 2\nu\pi/10$ ($\mu + \nu$ even)	(3b)
$\lambda = \arctan 2$	$\theta = 2\mu\pi/10$	$\phi = 2\nu\pi/10$ ($\mu + \nu$ odd)	(3c)
$\lambda = \pi/2$	θ = arbitrary	$\phi = (2\nu+1)\pi/10$	(3d)

for all $0 < \mu < 9$, $0 < \nu < 9$. One way to represent this structure is to consider a fixed value of θ and plot the (y,z) values that occur. In effect this is a projection along the great circle $\lambda = 0$ (the ten points of equation 3a); λ appears as a radial coordinate, and ϕ is a polar angle. For the choices $\theta = (2\mu+1)\pi/10$ there is only a point at the origin; but for $\theta = 2\mu\pi/10$ the ten values given by 3b and 3c form a star of five points as shown in Figure 1. The dark lies connect neighboring points in the "plane " of constant θ; dashed lines indicate neighbors in other planes. The ten points of Eq. 3d are difficult to fit in this scheme: the value of θ is undetermined, which means they are on all planes at once (the usual Mercator distortion of the poles). These points are indicated on the figure as the outer ring of particles. The stars for successive values of θ alternate in orientation between the one shown and its inversion.

We will wish to characterize disordered versions of this structure. This can be done by means of the Voronoi construction. Any four points in four-dimensional space determine a hyperplane, and a sphere with center on the

Figure 1: A projection of the {3,3,5} polytope. These are the y and z coordinates of all points for which $\mu = 1$ ($\theta = \pi/5$).

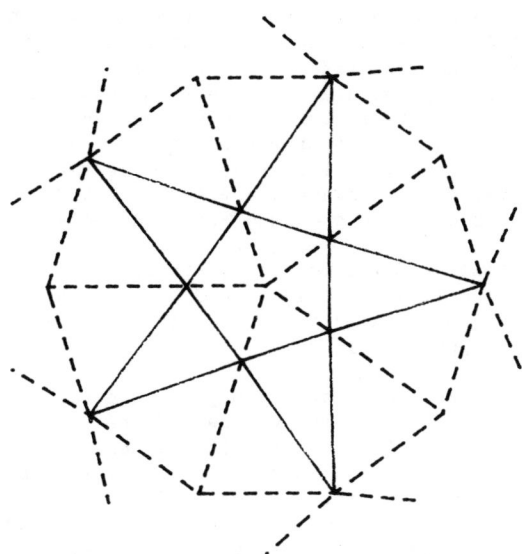

hyperplane. Given any set of points $\{r_i\}$, we will say that a subset of four of these form a neighbor tetrahedron if no other point in $\{r_i\}$ is closer to their insphere center than they are. Particles that belong to the same neighbor tetrahedron share a face of their Wigner-Seitz cells. The set of all neighbor tetrahedra spans the hypersphere surface: every point belongs to the interior of one tetrahedron, or the interface of two, or the edge of several, or is one of the points $\{r_i\}$.

The {3,3,5} polytope has a simple topology: every edge of a neighbor tetrahedron is shared by five others. Displacing the particles of the {3,3,5} polytope slightly does not change this characterization; but sufficiently disorganized sets of points can have other topologies, in which some number of neighbor tetrahedra other than five share an edge. These defective edges carry a conserved topological charge, so that a six-edge coming into a particle must be balanced by a six-edge departing, or two four-edges, or various other combinations[2].

Applying these ideas to Euclidean three dimensional space reveals some anomalies. The octahedron whose vertices are (0,0,0), (0,0,1), (.5,0,.5), (-.5,0,.5), (0,.5,.5), and (0,-.5,.5) must be resolved somehow into four neighbor tetrahedra, with the consequence that particles that are not nearest neighbors become "neighbors" in the sense of the Voronoi construction; the construction is ambiguous but on the average every particle has fourteen neighbors, and there is a high density of defect lines present.

Isolated Defects

Isolated defects can be constructed and studied in curved three dimensional space; these may be regarded as the low energy excitations of this problem. In view of the complexity of Euclidean three dimensions, it is clearly impractical to attempt to carry out the same program. As a simple example, we can construct a single 6-line wrapped around the equator of the hypersphere by letting μ range from 0 to 11 and letting θ be stepped in multiples of π/12 instead of π/10 (so that equation 3a reads θ = (2μ+1)π/12, etc). There are now 142 (=10 + 12 + 10 × 12) particles on the hyperspherical surface. The resulting structure differs from the picture above in that it is now a six-pointed figure (appropriately, since this defect is the "star" of David Nelson's theory[2]). This structure is not mechanically stable: there are unbalanced forces acting on most of the particles. But relaxing the structure by making small random motions that decrease the energy yields a stable structure with the same topology (in effect, this step adjusts the values of λ to relieve some of the crowding about the defect line).

By similar mutations of Eq. 3 we can construct the four-line (98 particles), the seven-line (164 particles), and linked configurations of two defect lines (by mutating both the θ and φ coordinates): 6 and 6 (168 particles), 6 and 4 (116 particles), etc.

Another class of imperfections in a crystal are vacancies and interstitial particles. From the point of view of the topological defect theory these are small closed clusters of defect lines. For example, by inserting an extra particle into the {3,3,5} structure and then annealing, the rather simple structure shown in figure 2 resulted. In this figure the solid lines are 6-lines and dashed lines are 4-lines. Only the bonds indicated changed topology; the other 109 particles had topologically icosahedral environment. The figure has three-fold symmetry about the vertical axis, with alternating triangular and trapezoidal wings.

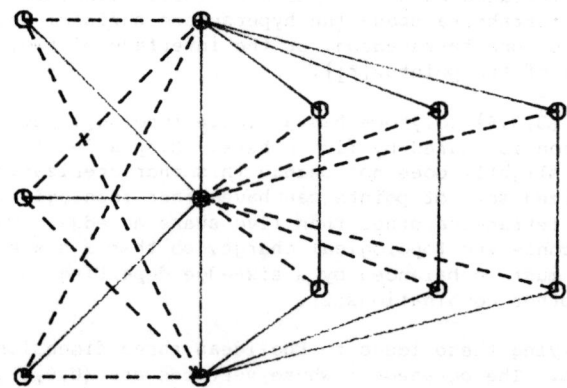

Figure 2. The topology of an interstitial.

For all these structures the energy

$$U = (1/2) \sum_i \sum_{\substack{j \\ i \neq j}} V(r_{ij}) \qquad (4)$$

was calculated. The results are given in table 1. An obvious source of the large variation is the different densities (all studies used the same sphere radius). The results have been rescaled to constant density by multiplying the energy of an N particle system by $(120/N)^4$: the density is greater by $N/120$ so r must be decreased by $(120/N)^{1/3}$; raising this to the 12th power gives the energy rescaling. The Voronoi construction was performed to determine the number of bonds; the rescaled energy per bond is also given.

The first two entries in table 1 are the icosahedral crystal and the glass which was reported previously (figures 2b and 3a of [3]). The energy of $\{3,3,5\}$ is almost exactly 4 per bond because all neighbor bonds are of unit length, and further neighbors contribute very little (see equation 1).

The 6-line, the 4-line, and the two linked 6-lines are stable against small random motions, but the 7-line and the linked 6- and 4-lines are not. Apparently both of the structures of linked defect lines are globally unstable.

The linked 6- and 4-line have 116 particles (8+12+8×12); an alternate structure with the same number of particles can be formed by removing 4 maximally separated particles from the $\{3,3,5\}$. Unfortunately this breaks the symmetry, because the set does not form a perfect tetrahedron. However, the energy is significantly lower, showing that the two lines must somehow merge and annihilate.

Table 1. Energy of defect structures

N	energy U	rescaled energies U	U/N	U/bond	comment
120	2904	2904	24.20	4.034	The $\{3,3,5\}$ itself
120	3148	3148	26.23	4.354	a glass (ref. 3)
119	2841	2938	24.69	4.109	best vacancy
121	3192	3088	25.51	4.212	best interstitial
116	3767	4314	37.19	6.000	6 line + 4 line
116	2680	3069	26.46		four vacancies
142	7933	4046	28.49	4.694	6 line
164	22478	6443	39.29	6.267	7 line
168	19028	4953	29.48	4.799	two 6 lines
168	17945	4671	27.81	4.423	best random start with 168 particles

REFERENCES

1) M. Rubinstein and D. R. Nelson, Phys. Rev. B <u>28</u>, 6377 (1983).
2) D. R. Nelson, Phys. Rev. B <u>28</u>, 5515 (1983)
3) J. P. Straley, Phys. Rev. B to be published
4) H. M. S. Coxeter, <u>Regular Polytopes</u> (Dover, New York, 1973).
5) J. P. Hansen, Phys. Rev. A <u>2</u>, 221 (1970).

This work was supported by National Science Foundation through Grant DMR-82-44366.

MICROSCOPIC STRUCTURE OF TWO-DIMENSIONAL ELECTRON GLASS*

R. K. Kalia and P. Vashishta
Materials Science and Technology Division
Argonne National Laboratory, Argonne, IL 60439

ABSTRACT

Structural properties of a classical two-dimensional system of electrons on a corrugated surface are investigated by the molecular dynamics method. Rapid quenching of the electron liquid leads to glass formation. Direct evidence for two-level states is obtained in the electron glass and the transitions between them are found to arise from displacements of a small number of spatially localized electrons.

Understanding the inherent structure of amorphous systems is one of the most challenging problems in condensed-matter physics. The linear specific heat and the quadratic temperature dependence of thermal conductivity at low temperatures provide clues to the hidden structure of glasses [1]. It has been postulated that these features arise from a statistical distribution of two-level states [2]. In this paper we discuss the inherent structure of a two-dimensional glass, providing direct evidence for the existence of two-level states from computer-simulation studies.

Monoatomic species on smooth surfaces do not form glasses even under very rapid quenching. Two-dimensional glasses are formed by quenching only in mixtures of particles of different sizes or monoatomic systems on curved surfaces [3]. We propose a third method which consists of particles on certain commensurate corrugated surfaces; for example, a classical electron glass formed on a sinusoidal corrugation. The corrugated surface can be realized experimentally by etching grooves on a metallic plate using photo or electron lithography. Immersing this plate in liquid He and applying a suitable voltage creates a corrugated potential for the electron monolayer on the He surface. The Hamiltonian of this system can be written as

---

Work supported by the U.S. Department of Energy.

$$H = \sum_{i=1}^{N} \frac{P_i^2}{2m} + \Phi , \tag{1}$$

where

$$\Phi = \sum_{i<j=1}^{N} \frac{e^2}{r_{ij}} - U_b - U_o \sum_{i=1}^{N} \sin kx_i . \tag{2}$$

In equation 2, U_b is the contribution of a uniform, positive background and the last term describes the effect of the corrugated surface. The wave vector, k, is chosen so that the ground state is a triangular lattice of electrons occupying alternate (either odd or even) corrugation minima. This doubly degenerate ground-state configuration is shown in figure 1. Molecular dynamics calculations were performed at a constant number density ($\rho = 1.477 \times 10^8$ cm^{-2}) and corrugation height ($U_o = 1$ K) using periodic boundary conditions and Ewald's summation for the long-range Coulomb interaction [4].

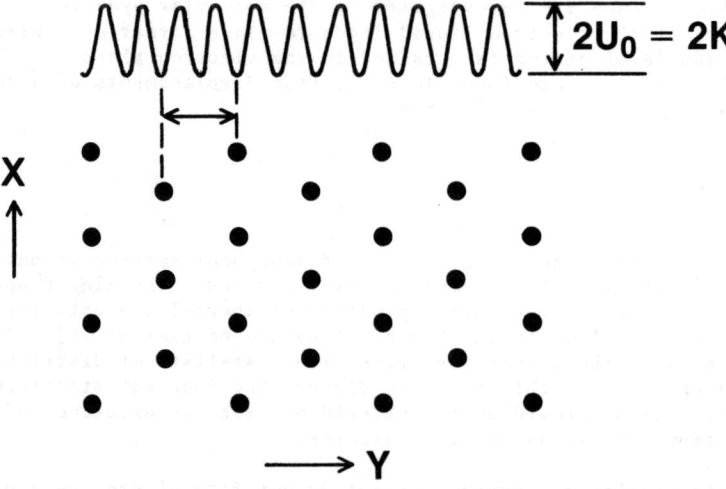

Figure 1. Ground-state configuration of electrons on a linear, periodic corrugation of wavelength which is half the spacing between electrons on a triangular lattice.

Figure 2 shows the total internal energy per particle as a function of temperature when the system is heated monotonically from the triangular lattice. The solid and open circles represent the solid and liquid phases, respectively. The onset of diffusivity and an abrupt change in the internal energy around $T_m = 0.55$ K indicate a first-order melting transition. Above T_m, both even and odd corrugation valleys are uniformly occupied in the liquid state.

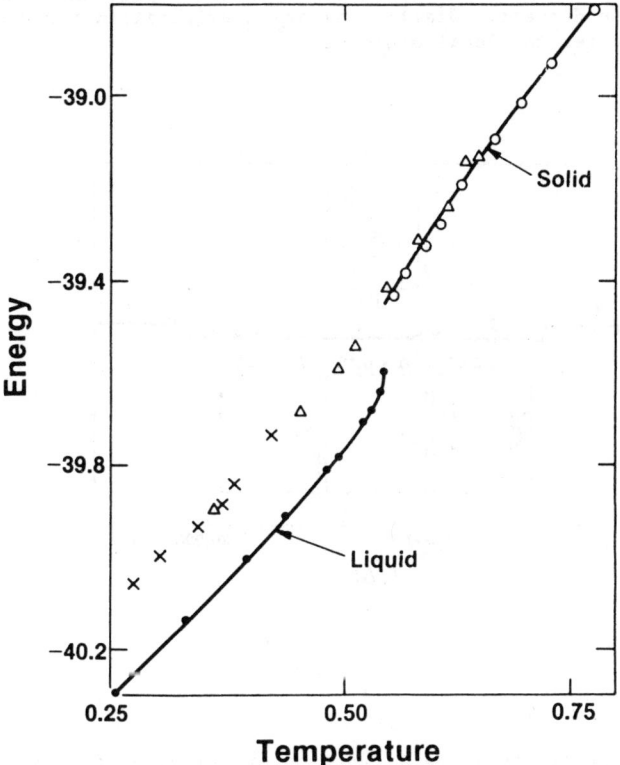

<u>Figure 2</u>. Temperature variation of the energy per particle in degrees Kelvin. The continuous lines are drawn to guide the eye.

The glassy states were obtained by rapidly cooling the liquid states and then allowing the systems to run for 30,000-40,000 time steps (MD time step, $\Delta t = 2.5 \times 10^{-12}$ sec) after equilibration. These states, shown by crosses in figure 2, did not exhibit any diffusion. The triangles represent the states obtained by monotonically heating a glassy state. The electron glass exhibits continuous melting around T_m.

The statistical description of a solid or glass can be separated into mechanically stable packings of particles corresponding to local potential-energy minima and the vibrational motion about these stable positions [5]. It is possible to enumerate the local minima of a state by a numerical procedure called the steepest-descent quench. This involves moving "downhill" from the starting point in the configuration space along a steepest-descent direction which connects the initial configuration to the minimum by the multi-dimensional equation

$$\frac{\partial R}{\partial s} = - \nabla \Phi(R) , \qquad (3)$$

where R is a vector in the 2N-dimensional configuration space and s is a "virtual" time for the descent. Starting at any configuration R(s = o), the solution R(s → ∞) locates the local minimum.

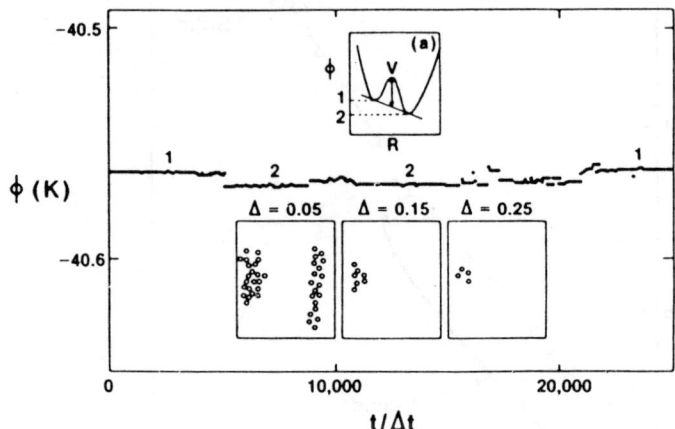

Figure 3. Potential-energy minima as a function of molecular-dynamics time at which the electron glass at T = 0.22 K is quenched. Note, the sequence of bi-level states in φ vs t corresponds to the two-level states in the configuration space shown schematically in the inset (a). The other inset is a snapshot of particles whose movements by an amount greater than Δ = 0.05, 0.15, or 0.25 induce the transitions between levels 1 and 2 around t = 5100 Δt.

An ensemble of glass configurations from the molecular dynamics run are quenched along the steepest-descent directions at closely spaced times to obtain the corresponding Φ_{min}. Figure 3 shows $\phi = N^{-1}\Phi_{min}(t)$ as a function of time at which the steepest-descent quench is applied to a glassy state at T = 0.22 K. For the first 5,100 time steps, the minimum potential energy of configurations is designated by level 1. The system then finds itself in the vicinity of another level, 2, and for the next 3,600 steps level 2 remains as the minimum potential energy of the glassy state. Around 8,800 Δt, the system has a potential-energy minimum which is between levels 1 and 2. However, after another 2,000 time steps the system returns to level 2 and remains there for the next 10,000 steps. During this time, the system attempts short transitions between levels 1 and 2 and then around 21,000 Δt it successfully jumps to level 2. These results clearly manifest a distribution of two-level states in the electron glass, as shown schematically in figure 3a.

The inherent structure of the electron glass is analogous to the two-level states first postulated by Anderson, Halperin, and Varma, and Phillips [2]. According to these authors, a large number of modes having small

$\Delta\phi$ (= $\phi_1 - \phi_2$) may be inaccessible because of large energy barriers between them or because the transitions between them require the cooperative motion of too many particles. For the electron glass at T = 0.22 K, we have calculated the change in the position of particles just before and after the transition from level 1 to 2 at t = 5,180 time step. Denoting $\delta_i = |\vec{r}_{i1} - \vec{r}_{i2}|/d$, where $\vec{r}_{i1}$ is the position of the ith particle in level 1 and d the lattice spacing, we show in the inset to figure 3 a snapshot of only those particles which have $\delta_i \geq \Delta = 0.05$, 0.15, and 0.25. For $\Delta = 0.05$ we find two regions, each containing a few particles. However, when Δ increases to 0.15 or 0.25 only a few particles confined to a small region induce the transition between two-level states. This confirms the postulate of Anderson et al. and Phillips [2].

REFERENCES

1) Pohl, R. O.: <u>Amorphous Solids, Low-Temperature Properties</u>, ed. Phillips, W. A.: (Springer-Verlag, Berlin) (1981), p. 27.

2) Anderson, P. W., Halperin, B. I., and Varma, C. M.: Philos. Mag. <u>25</u>, 1 (1972); Phillips, W. A.: J. Low Temp. Phys. <u>7</u>, 351 (1972).

3) Nelson, D. R.: <u>Topological Disorder in Condensed Matter</u>, eds. Yonezawa, F. and Ninomiya, T. (Springer-Verlag, Berlin) (1983), p. 164.

4) Vashishta, P. and Kalia, R. K.: <u>Melting, Localization, and Chaos</u>, ed. Kalia, R. K. and Vashishta, P. (North-Holland, New York) (1982), p. 43.

5) Stillinger, F. H. and Weber, T. A.: Phys. Rev. A<u>25</u>, 978 (1982); see also Phys. Rev. A<u>28</u>, 2408 (1983).

PHASE TRANSITIONS IN CRYSTALLINE POLYMERS

P. L. Taylor
Department of Physics
Case Western Reserve University
Cleveland, Ohio 44106

ABSTRACT

The crystal structure of a polymeric system is determined by a balance between intrachain and interchain forces. The constraint of a crystalline environment generally means that a chain is free only to rotate or to coil, while the packing of the chains will be determined by the long-range and short-range forces between them. Three simple examples are discussed of crystalline polymers that exhibit phase transitions under the influence of stress, heat, or applied electric fields.

INTRODUCTION

The theory of phase transitions now embraces a vast number of techniques that may be applied to a wide variety of systems, and a great deal is now felt to be understood about the way in which a change of phase occurs. The most difficult models to treat are those in which the energy eigenvalue spectrum is unbounded, as is the case for most models of the transition from a liquid to a solid or vapor. Much simplification is achieved when the positions or velocities of the particles are confined to a finite range of discrete values, and the most simplified models of all describe a system such as the Ising ferromagnet, in which the state of a large system is characterized by the values of a set of variables confined to take on one from some small number of possible choices. Some such models may even be solved exactly for some properties such as specific heat or entropy.

Crystal-crystal phase transitions in polymers unfortunately do not fall in the category of problems that can be accurately modeled and exactly solved; that would be an unrealistic hope. They do, however, have some advantages that make them an attractive object for study. To see this, let us look at the construction of a simple polymer, such as polyethylene (figure 1). This material has a backbone of carbon atoms that are covalently bonded. The forces involved are very strong, and involve energies of the order of an electron volt. We know from everyday experience that if we drop a polyethylene object into boiling water, then it will soften but not decompose. Much higher temperatures are necessary to split the polyethlene chain into its constituent ethylene units, as is borne witness by the reluctance of polyethylene to burn when exposed to a match. We can thus assume that at modest temperatures each carbon-carbon bond remains uncut.

This does not leave a lot of options open for phase transitions between crystal phases. All that is left is the possibility that the chain may crinkle or rotate in some way, or that the arrangement of the chains to form a two-dimensional lattice in the plane perpendicular to their length may be altered. Phase transitions in crystalline polymers thus bear some similarity to martensitic transformations in metals, in that if two particular atoms are near neighbors in one phase then they will remain neighbors in the other phase, and any numbering of atoms in a chain will remain valid through the phase transition.

Not all polymers, of course, are obliging enough to form good crystals that can be described in this way. Poly(vinyl chloride), or PVC, is a plastic produced in vast quantities for all sorts of domestic uses, and is nearly always completely amorphous. The reason for this is the fact that every second carbon atom along the chain has a chlorine atom (figure 2) in place of one of the hydrogens that are the only atoms attached to the carbons in polyethylene. When PVC is manufactured, the vinyl chloride units join on to each other to make a chain in such a way that each chlorine atom is almost equally likely to be pointing up out of the plane

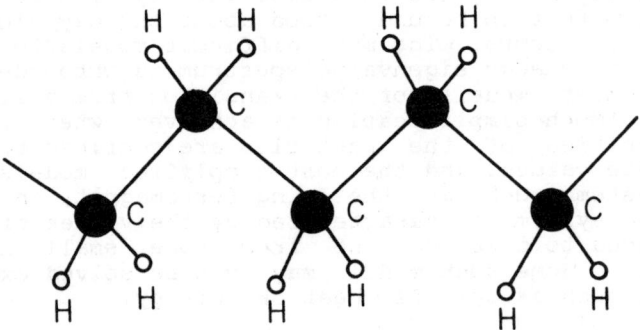

Figure 1. Structure of polyethylene.

of the nearest three carbons as it is to be pointing downwards. The jargon for this phenomenon is to say that PVC is usually <u>atactic</u>. The consequence of this disorder is that PVC has little tendency to form an ordered crystalline array.

If we were to replace <u>both</u> the hydrogen atoms on every second carbon by a chlorine we would obtain poly(vinylidene chloride), $PVCl_2$, which would be much easier to crystallize. If we were to use fluorine instead of the larger chlorine atoms, we would have poly(vinylidene fluoride) or PVF_2 (figure 3). This is a truly fascinating material, possessing all sorts of possibilities for use as a piezoelectric or pyroelectric sensor. Even this molecule, however, which is much more symmetric than PVC, is not easily made into single crystals. Because most polymers are synthesized from monomers in the liquid or gaseous phases, the long chain tends to get knotted and tangled with itself and other

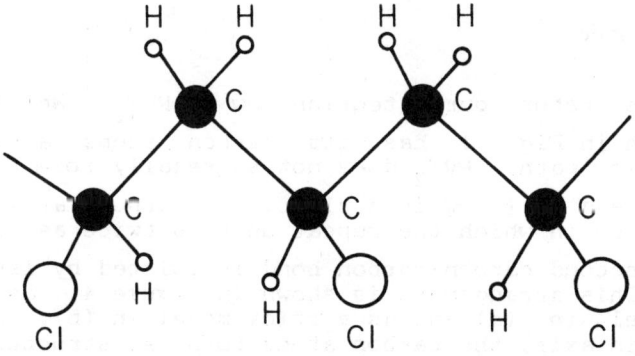

<u>Figure 2</u>. Structure of the disordered form of poly(vinyl chloride).

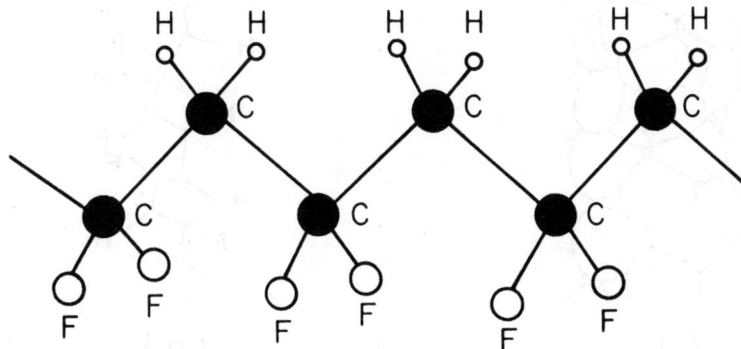

<u>Figure 3</u>. Structure of the beta phase of PVF_2.

chains, and to defy crystallization to produce a material without appreciable amorphous content. One can make an analogy with trying to cook spaghetti in a big saucepan (perhaps starting with a single spaghetto, one kilometer long) and then hoping to shake the pan until it settles down in an ordered array.

There is one way, incidentally, in which to produce good single crystals of some polymers. The secret is to cheat. One starts by making a crystal of monomer, and this is no problem, as the molecules involved are small. If this monomer is sufficiently unstable and energetic, perhaps being full of triple bonds between carbon atoms, then some small perturbation, such as gamma radiation or a rise in temperature, may be sufficient to start a polymerization reaction that will proceed in a straight line down the crystalline array. Like a zip fastener, it joins together the atoms by changing a triple bond to a double one, and using the spare bond to link neighbors. This process is known as solid-state polymerization.

Polyvinylidene Fluoride

Let us return our attention now to PVF_2. While the polyethylene shown in Fig. 1 has its carbon atoms arranged in a planar zig-zig path, PVF_2 does not so readily form this type of structure. Insead of lying in a plane, the carbon atoms tend to form a structure in which the repeat unit is twice as long, and in which every second carbon-carbon bond is twisted by 120° from its planar form. This arrangement is shown in figure 4, as a space-filling model in (a) and as a stick model in (b). When viewed along the chain axis, the carbon atoms form a structure looking

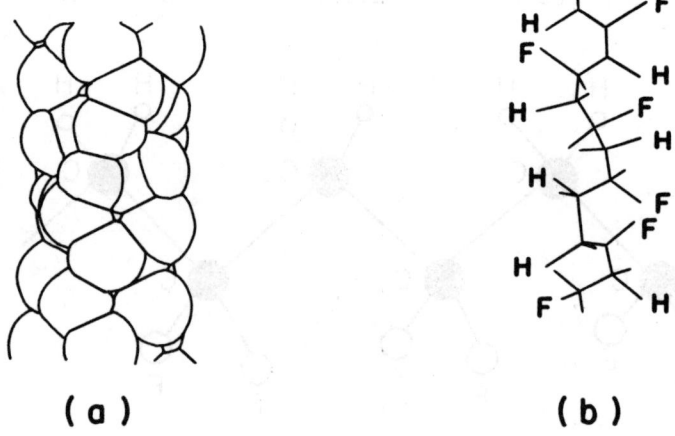

(a)   (b)

<u>Figure 4</u>. Space filling (a) and stick model (b) of alpha-PVF_2.

like an array of bow ties. Figure 5 shows this crystalline phase, which is known as the alpha phase, in a drawing in which the hydrogen atoms have been suppressed, and the carbon atoms reduced to points. The circles show fluorine atoms, and it is clear from this picture that the alpha phase is an antiferroelectric. The fluorine atoms each carry a considerable negative charge, stolen from the carbon atoms in an attempt to fill fluorine's unfilled outer shell. The asymmetric nature of the molecule gives each individual chain a dipole moment, but the orthorhombic crystal contains chains whose moments are opposed.

The fact that the carbon backbone of alpha-PVF_2 is coiled while that of the structure shown in figure 3 is flat suggests that it might be possible to transform one into the other by pulling hard enough on the ends of the chains. This is indeed possible, and yields the phase known as beta-PVF_2. In this structure, the chains pack in such a way as to have their dipole moments aligned, and it is thus a ferroelectric. Unlike ceramic ferroelectrics, in which the moment arises from an instability in the phonon spectrum of a non-polar material, beta-PVF_2 consists of a flexible chain composed of units of fixed dipole moment of about 7×10^{-30} coulomb-meters. The great technological interest in beta-PVF_2 arises from the fact that one may make transducers and other devices of large area from it. It may be produced in complex shapes, is easily cut without damage, and is light and flexible. It requires no complex mounting process, is chemically stable and mechanically strong, and can handle large electrical power. Its

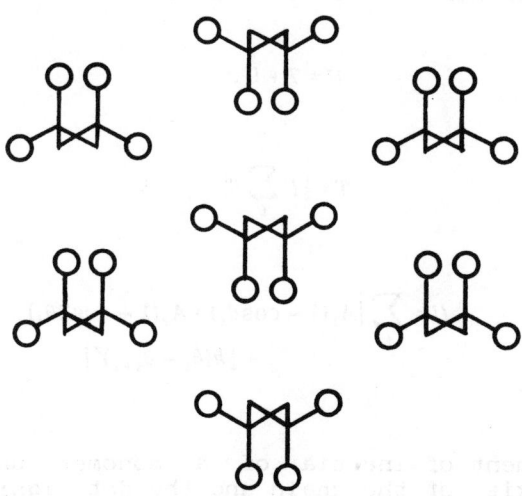

Figure 5. Orthorhombic cell of alpha-PVF_2.

density is sufficiently low that it has a better acoustical match to water than most other ferroelectrics. Its piezoelectric properties make it useful in hydrophones, loudspeakers and optical modulators, while its pyroelectricity lends itself to use in infrared detectors.

As produced by drawing from the more stable alpha phase, a film of beta-phase material will consist of randomly oriented domains. In order to obtain useful aligned material, this sample is exposed to large electric fields at modestly high temperatures, a process known as poling. To understand the process by which the chains rotate about their axes to align themselves with the applied field has been a challenging task in theoretical polymer physics.

The first model to be studied was one introduced by Aslaksen [1], in which the polymer chain rotates by 180° about its axis. The second model followed a more recent suggestion by Kepler and Anderson [2], who noticed that a rotation by 60° might also occur. Both these models led to results that were at great variance with the experimental results as known at the time. Fortunately, this uncomfortable situation was relieved by some more careful experimentation, which showed that a theory based on one of these models had in fact predicted the observed relaxation time for the poling process.

The first model considered the situation shown in figure 6, in which a domain wall separates oppositely polarized regions of a crystallite. A phenomenological Hamiltonian may then be proposed of the form

$$H = T + U,$$

where

$$T = \tfrac{1}{2} I \sum_i \dot{\theta}_i^2 \qquad (1)$$

and

$$U = \sum_i [A_1(1 - \cos\theta_i) + A_2(1 - \cos 2\theta_i) + \tfrac{1}{2} k (\theta_i - \theta_{i+1})^2] \qquad (2)$$

where I is the moment of inertia of a monomer unit about the center-of-mass axis of the chain and the dot signifies differentiation with respect to time. The first two terms in U represent the combined influence of the local crystalline order (that is, the interchain potential) and an applied electric field. They constitute a potential having minima at $\theta = 0$ and π providing

$4A_2 > |A_1|$. In practice, even with large applied electric fields, this condition is satisfied. The last term in U represents the torsional rigidity of the chain. The interchain potential for a monomer unit of a chain at the center of the beta-phase unit cell is shown as a function of the rotational angle θ about its axis in figure 7.

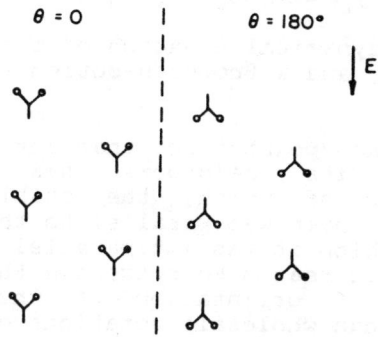

Figure 6. At the boundary between oppositely polarized regions some chains are in a neutral environment for which the energies of either of two opposite orientations are approximately equal.

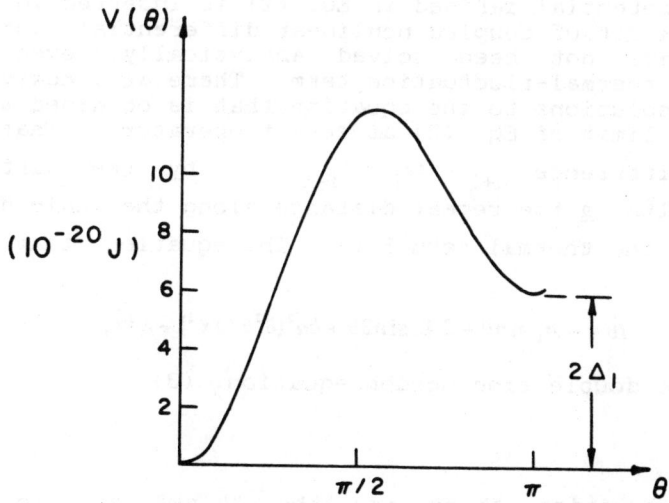

Figure 7. Interchain contribution to the potential energy of a monomer unit of PVF_2 as a function of angle of rotation about its chain axis. The unit is located at the center of the orthorhombic unit cell of the beta phase. All neighboring chains are aligned at $\theta = 0$.

The time variation of J_i, the angular momentum of the ith unit, defined as

$$J_i = I\dot{\theta}_i \qquad (3)$$

obeys a Langevin equation of the form

$$\dot{J}_i = -\partial U/\partial\theta_i - \lambda J_i + F_i(t). \qquad (4)$$

This equation is the dynamical equation of the system supplemented by a damping term $-\lambda J_i$ and a Brownian-motion term $F_i(t)$.

Aslaksen [1] had pointed out that any polymer chain in the crystal field of its neighbors has two locally stable orientations, in one of which, the stable orientation, the polarization of the chain was parallel to that of its neighbors, and in the other of which it was antiparallel to the neighbors' orientation. For this reason he suggested that poling involved a possible reversal of orientation of the polarization of a crystallite, rather than wholesale rotations of crystallites. The poling field would then lower the energy of the metastable antiparallel orientation of the chain polarization in such a way as eventually to reverse the orientations of all polymer chains within a crystallite. The advantage of this picture is that the topological structure of a polymer sample is preserved, and large-scale motions of polymer chains, corresponding to changes of orientation of the local crystal axes, are avoided.

When the potential defined in Eq. (2) is inserted in Eq. (4) one obtains a set of coupled nonlinear differential equations of motion which have not been solved analytically, even in the absence of the thermal-fluctuation term. There are, however, some known special solutions to the equation that is obtained by taking the continuum limit of Eq. (2) at zero temperature. That is, one replaces the difference $\theta_{i+1} - 2\theta_i + \theta_{i-1}$ by the differential $a^2\partial^2\theta/\partial x^2$ with $\underline{a}$ the repeat distance along the chain direction x, and ignores the thermal term $F(t)$. The equation that results,

$$I\ddot{\theta} = -A_1\sin\theta - 2A_2\sin2\theta + ka^2(\partial^2\theta/\partial x^2) - \lambda I\dot{\theta}, \qquad (5)$$

is known as the double sine-Gordon equation. (3)

To this equation there are the "kink" solutions, which describe the motion of the boundary between two regions of the chain, in one of which $\theta \simeq 0$ and in the other of which $\theta \simeq \pi$. In the present case the potential energy is lower for $\theta \simeq \pi$ and so this region advances at the expense of the region in which $\theta \simeq 0$. This is illustrated in figure 8.

The approximation that yields the double sine-Gordon equation has a number of obvious weaknesses. One knows, for example, that the poling process is temperature dependent, and so the exclusion of the Brownian-motion term F(t) from Eq. (4) allows only a zero--temperature result to be predicted. One must also expect the discrete nature of the polymer chain to play a significant role in the motion of a kink of polarization.

In view of these inadequacies of the analytical approach, a series of computer experiments were performed in which the discrete nature of the chain and the thermal fluctuations were taken into account. A chain of 40 monomer units was chosen for study, in accord with experimental observations of lamellar thickness (2). The equations of motion, Eqs. (4) were solved by linearization and direct integration over a small time increment Δt. The thermal forces were included by adding an impulse of fixed magnitude $(2\lambda I k_B T / \Delta t)^{\frac{1}{2}}$ but of random sign during each time increment. Because of the random nature of the force, all the finite-temperature experiments were performed repeatedly and average values taken.

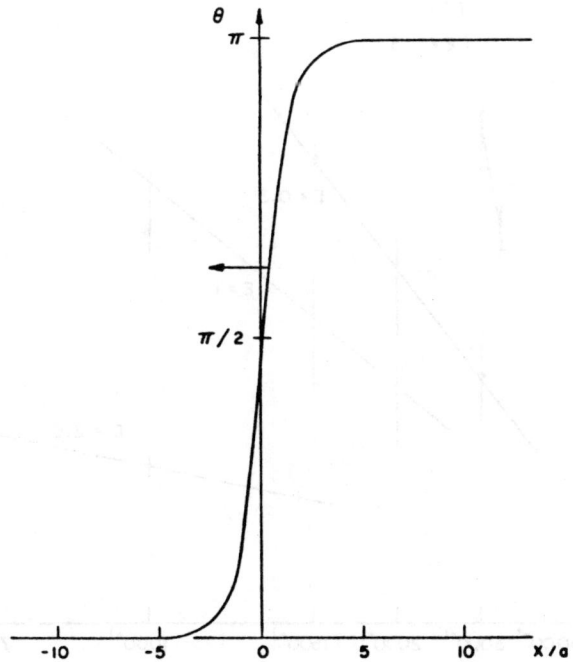

Figure 8. Solution of the double sine-Gordon equation represents a kink traveling to the left.

A computer experiment was then performed to determine the waiting time t_w that on average elapses between application of the field E and passage of a kink into the crystallite along a given chain. The results of this study are shown in figure 9, in which t_w is plotted as a function of inverse temperature for various fields. The approximately linear behavior suggests that a picture of thermally activated kink creation is valid, with

$$t_w = A e^{B/T}.$$

The "attempt frequency" A^{-1}, found from the common extrapolations of the lines to infinite temperature, yields a value of about 7 THz, which is to be compared with the figure of 8.4 THz obtained by multiplying twice the librational frequency at

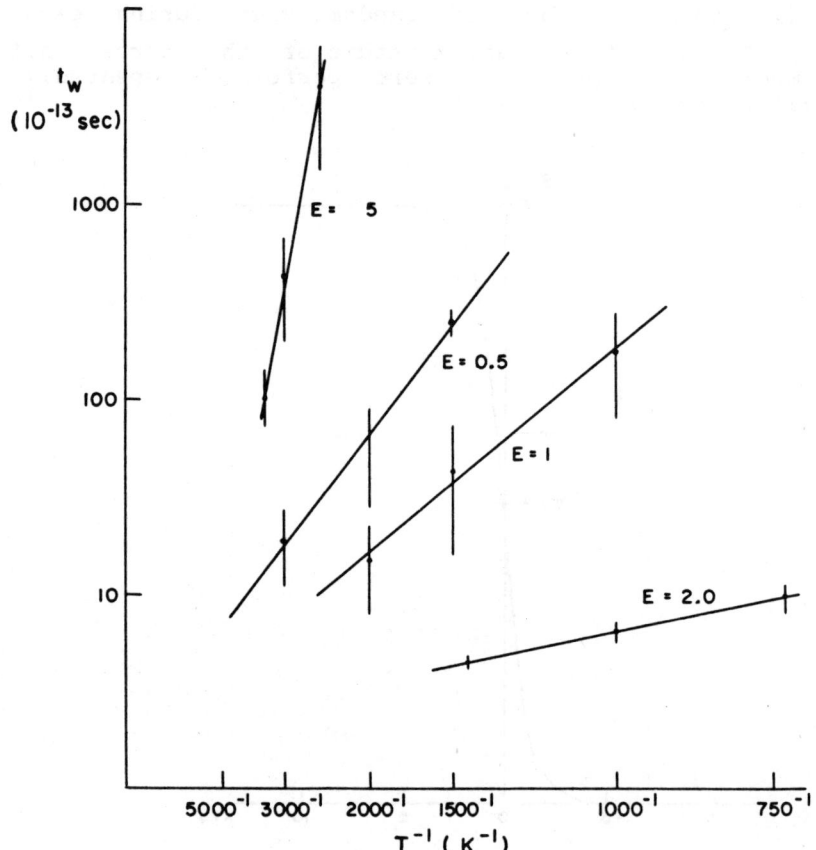

Figure 9. Logarithm of the average waiting time t_w for creation of a 180° kink is shown as as function of T^{-1} for various fields E in units of 10 GV m^{-1}.

zero wave number of 2.1 THz by a further factor of 2 to allow for the two ends of the chain. The activation energy Bk_B, found from the slope of these lines, can be interpreted satisfactorily in terms of the contributions of the various terms in the Hamiltonian given in Eqs. (1) and (2).

The average waiting time for a chain to reverse its electric polarization under typical poling conditions (373 K, 500 MV m^{-1}) and in a neutral crystal environment is predicted by this model to be of the order of 7×10^4 sec. The theoretical poling time for the entire sample will clearly be much larger than this, and of the order of days or weeks. Because poling had been observed to occur in periods of a few minutes, there was clearly some inadequacy in the model.

The clue to the alternative mechanism for poling came from the orthorhombic nature of PVF_2. The closeness of the structure to a hexagonal form led Kepler and Anderson [2] to make an ingenious suggestion of an alternative mechanism of poling, namely that a rotation of the chains through $60°$ rather than $180°$ plus a small distortion of the lattice might be the correct model. This possibility is illustrated in figure 10, in which the boundary between such twinned regions is shown. Some support for such a mechanism has recently been given in the form of infrared studies in which the optical properties of a PVF_2 surface were shown to be modified by poling [5]. A simple reversal of chain orientation through $180°$ would presumably have left these unchanged.

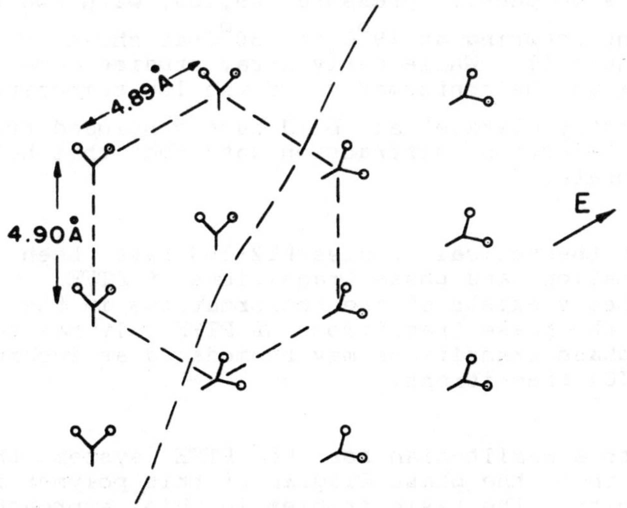

Figure 10. Boundary between regions of beta-phase PVF_2 differing in direction of polarization by $60°$.

In this model the potential barrier to rotation was greatly reduced, and poling proceeded much more rapidly than in the previous case. In fact, the predicted poling time was now reduced to be of the order of microseconds to milliseconds [6]. We had escaped from the frying pan of having predicted a poling time too long by three orders of magnitude only to land in the fire of a prediction apparently too short by six orders of magnitude. It was thus a welcome resolution when Furukawa and Johnson [7], in a careful set of measurements of the poling process, concluded that relaxation to the poled state occurs in times of the order of microseconds. Subsequent analysis [8] showed that the detailed form of the growth of the polarization with time was matched quite closely by this model of propagating kinks. There is probably some sort of moral in this story: if you risk the laughter of your colleagues by publishing theories that differ from experiment by enormous factors, you may one day have the great pleasure of seeing the experimental results revised in your favor. On the other hand, you may not.

Polytetrafluorethylene

Teflon, or PTFE, is well known for its "non-stick" properties. Eggs, in principle, can be fried in a pan lined with PTFE and then removed with no adhering traces left behind. This same slipperiness that makes Teflon unwilling to bind to other substances shows up also in the ease with which the polymer chain can rotate about its crystalline axis. Early experiments [9] established that under atmospheric pressure PTFE exists in the form of helical chain molecules. At least three solid phases have been identified at atmospheric pressure [9,10], with two structural phase transitions occuring at $19^{\circ}C$ and $30^{\circ}C$ as shown in the phase diagram in figure 11. While early x-ray studies found a commensurate 13_6 helix as the conformation of the low-temperature phase (phase II), recently Clark et al. [11] have concluded from careful examination of electron diffraction data that this helix is, in fact, incommensurate.

A number of theoretical studies [12-16] have been conducted on the conformation and phase transitions of PTFE. However, no comprehensive theory exists of the conformations in the different phases or of the phase transitions of PTFE. It has been speculated that the phase transitions may be modeled as incommensurate--commensurate (IC) transitions.

One route to a Hamiltonian for the PTFE system that might yield insight into the phase diagram of this polymer is through molecular mechanics. The basic problem in this approach is the determination of universally applicable parameters characterizing the nonbonded interactions. Such parameters are never really universal and do not produce very accurate results, energy differences of 1 kcal mol^{-1} being barely significant in calculations using this method. (The barrier between the right- and left-handed helices in PTFE is estimated to be about 2 kcal mol^{-1}.)

This being the case, Banerjea [17] has formulated a model in which the most basic elements of the Teflon chain are present, but in which the mathematics is sufficiently tractable to permit solution of the relevant equations.

Teflon has a molecular structure similar to the polyethylene shown in Fig. 1, but with all the hydrogen atoms replaced by fluorines. A simple model of this system may be visualized as a linear array of canted "arrows" or "spins" of equal size, each free to rotate in a plane perpendicular to the line joining the arrows, as shown in figure 12. The angle that the jth arrow makes with a certain fixed direction, defined as the z-direction is ϕ_j. Each arrow interacts with its nearest neighbors with a potential energy of the form:

$$W(\phi_{n+1}, \phi_n) = -\cos(\phi_{n+1} - \phi_n - \alpha) \qquad (6)$$

Here α is a measure of the degree of natural cantedness of the system. In other words, to minimize this energy in the absence of other forces the arrows form a helix, and α is a measure of the

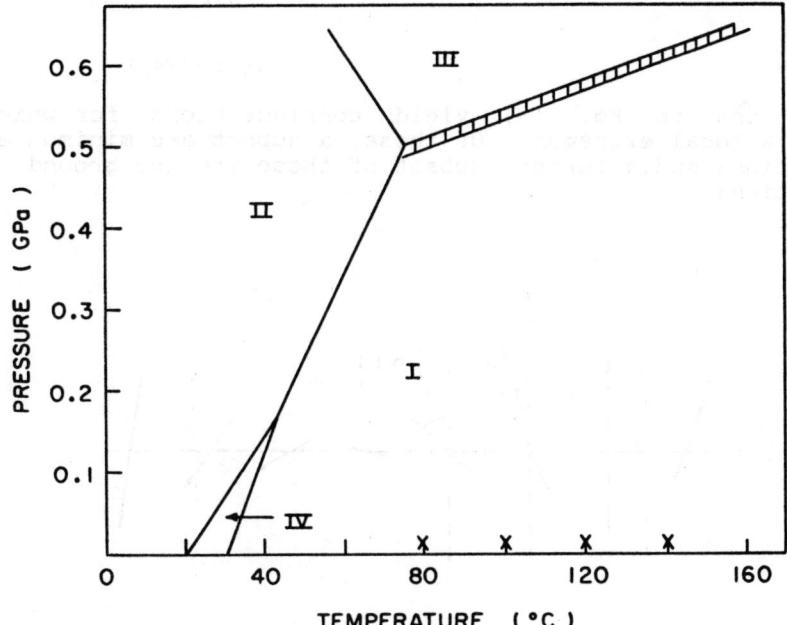

Figure 11. Phase diagram of Teflon.

pitch of that helix. These arrows, or CF_2 units, are also subject to an external symmetry-breaking field which represents the effect of interchain interactions. The potential energy of the jth arrow in the presence of the external field is given by:

$$V(\phi_j) = - \gamma \cos(2\phi_j) \qquad (7)$$

The constant γ characterizes the strength of the interchain forces, and hence the pressure. Under the influence of this pressure alone, i.e. with no nearest-neighbor interaction, the ground state configuration would have all the arrows aligned parallel or anti-parallel to the z-direction.

The total Hamiltonian for the array of arrows is then

$$H = - \sum_n \{\cos(\phi_{n+1} - \phi_n - \alpha) + \gamma \cos(2\phi_n)\} \qquad (8)$$

The conditions for a ground-state configuration may be found simply by minimizing the total energy with respect to the set of variables $\{\phi_n\}$. Differentiating the total Hamiltonian with respect to ϕ_n and setting the derivative equal to zero gives an infinite set of equations:

$$\sin(\phi_{n+1} - \phi_n - \alpha) - \sin(\phi_n - \phi_{n-1} - \alpha) \qquad (9)$$

$$= 2\gamma \sin(2\phi_n)$$

Solutions to Eq. (9) yield configurations for which the energy is a local extremum. Of these, a subset are minimum energy configurations and a further subset of these are the ground state configurations.

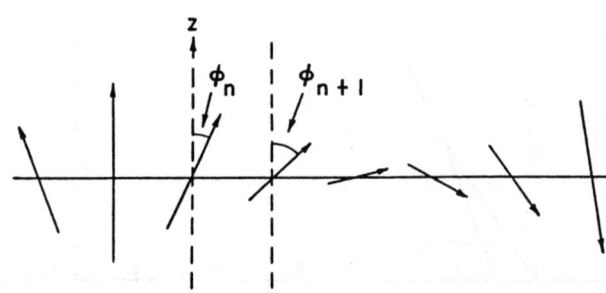

<u>Figure 12</u>. In the helical molecule of PTFE successive CF_2 units are oriented at angles ϕ to the crystal z axis.

The equations determining the condition for a ground state, Eq. (9), transform, in the continuum limit, to a non-linear second-order differential equation. It is quite common in dealing with second-order differential equations to reduce each to two coupled first-order differential equations. A similar operation can be carried out with the difference equations. In fact, such a procedure applied to the discrete sine-Gordon equation gives rise to the well-known and much-studied standard map, also referred to as the Taylor-Chirikov [18] map or the Frenkel-Kontorova map. In the case of Eq. (9) this reduction can be effected as follows.

A new variable s_n is defined through the equation

$$s_n = \sin(\phi_{n+1} - \phi_n - \alpha) \tag{10}$$

Then Eq. (9) can be rewritten in terms of ϕ_n and s_n as

$$s_n - s_{n-1} = 2\gamma \sin 2\phi_n \tag{11}$$

Together, Eqs. (10) and (11) form a pair of equations analogous to a pair of coupled first-order differential equations.

The variable ϕ is defined as modulo π and lies in the range $[0, \pi]$ while the variable s is restricted to lie in the range $[-1, +1]$ in order that the inverse sine be well defined. With these restrictions, Eqs. (10) and (11), which we rewrite here for convenience in the form

$$\phi_{n+1} = \phi_n + \alpha + \sin^{-1} s_n \tag{12}$$

$$s_{n+1} = s_n + 2\gamma \sin 2\phi_{n+1} \tag{13}$$

define a mapping of a part of the surface of a cylinder onto itself. This mapping, which can be expressed as

$$G(\phi_n, s_n) \to (\phi_{n+1}, s_{n+1}) \tag{14}$$

can be trivially shown to be area-preserving.

While no complete solution is possible, Banerjea was able to make a number of conjectures within the context of this simple model of PTFE. One of the most interesting ideas results from the observation that increasing pressure induces a commensurate phase to form at certain energies, but that there is an accompanying region of chaotic states of higher energy.

This is shown in figure 13, which is a map of s_n as a function of ϕ_n. In (a) the pressure is low, and the continuous lines are indicative of an incommensurate system, with all possible values of ϕ occuring, although for some energies these lines have broken up into rings, which are characteristic of commensurate states. At higher pressures, as shown in (b), the commensurate states are surrounded by a chaotic region in which s_n is not restricted to a finitely valued function of ϕ. One may speculate that these disordered states may play a crucial role in permitting rapid relaxation of one phase of PTFE to another as the temperature or pressure is varied.

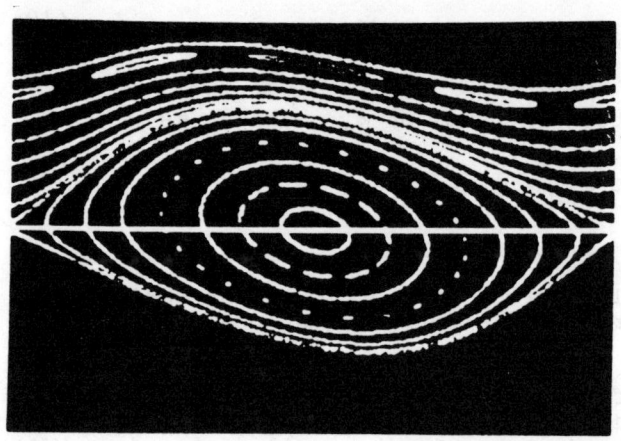

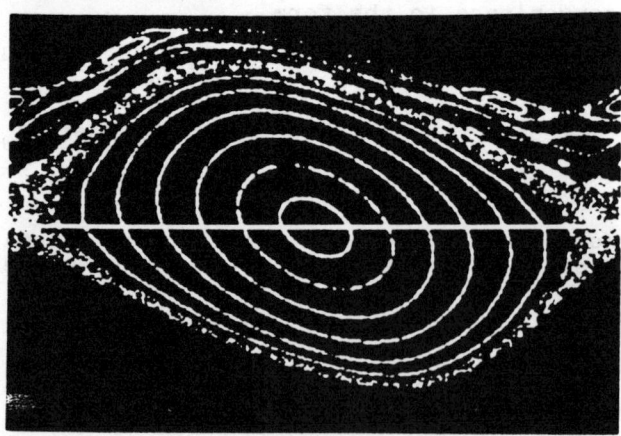

Figure 13. A plot of $s(\phi)$ shows how the angle between successive CF_2 units varies with the angle between a CF_2 and the crystal axis. The left-hand figure shows commensurate states as isolated rings or dots; the right-hand figure shows that at higher pressure chaotic states exist for which s is not a simple function of ϕ

Polybutylene Terephthalate

This almost-unpronounceable polymer is usually known as PBT, although some workers refer to it as PTMT, which is an abbreviation for the totally unpronounceable alternative name of polytetramethylene terephthalate. The repeat unit of the polymer chain consists of the terephthalate part, which is a benzene ring with a couple of carbon dioxides attached on opposite sides, and a linking butylene piece of connected CH_2 units, as shown in figure 14.

The interesting properties of PBT reside in the fact that the butylene part has two possible conformations that are very close in energy. One of these is an extended conformation, in which the carbon atoms form a planar zig-zag, while the other has the butylene segment slightly coiled up. Luckily (for the wearers of stretch jeans, many of which now incorporate this elastic polymer) the coiled conformation has the lower energy, and is thus the stable form. It takes only a modest stress, however (and modesty is not an unimportant concept in this area) to produce a transition to the extended form.

The crucial aspect of this transition from the coiled conformation, which is known as alpha-PBT, to the extended beta-PBT is that it appears to be totally reversible. On removal of the stress from the beta chain, the alpha chain is recovered. This property makes PBT almost unique among the simple polymers, and poses a challenge to theoretical analysis.

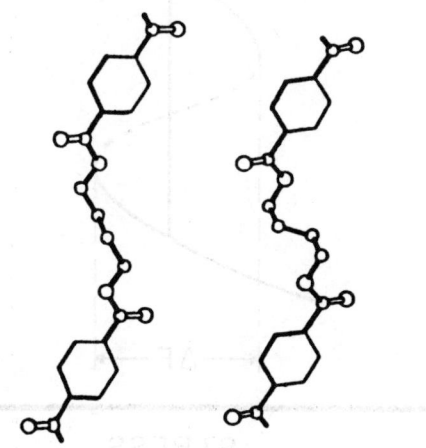

Figure 14. Chain conformation in the α(left) and β(right) forms of PBT.

The fact that this stretchable property of PBT does represent a true crystal-crystal phase transition has now been demonstrated by a number of experimental techniques. Perhaps the most convincing of these is the type of dynamical x-ray measurement now possible using synchrotron radiation. There one can see the disappearance of some diffraction spots and the appearance of others as the crystal is strained over periods of a few milliseconds, the transition being apparently complete at strains of about ten per cent.

Very little is yet known about the detailed mechanism by which this polymer expands and contracts, although some competing models are starting to be suggested. Vandana Datye [19] has studied the question of whether neighboring benzene rings in adjacent chains remain linked to each other (Model I), or whether one chain can translate along its axis like a worm in a tunnel (Model II). Preliminary results seem to indicate that a model of linked chains yields better predictions for the stress at the critical point than does the alternative; these calculations were only performed within a mean-field model, and so too great a reliance should not be placed on them.

Some support, however, for the validity of the mean-field model in this context comes from an examination of the width of the hysteresis loops found in the stress-strain curves for PBT.

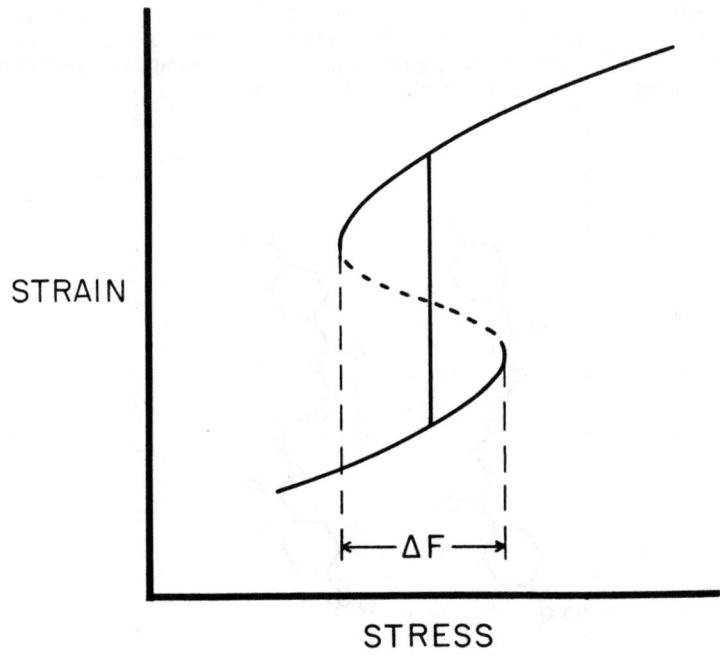

Figure 15. Hysteresis in the mean-field model of a stress-strain curve.

As one approaches the critical temperature T_c for the alpha-beta phase transition, the discontinuity in strain vanishes roughly as $(T_c-T)^{\beta'}$ with β' a positive number. Similarly, the elastic modulus (the derivative of stress with respect to strain) should vanish as $(T_c-T)^{\gamma'}$. The exact function describing stress as a function of strain is not expected to be analytic at the transition, and so it is not legitimate to extrapolate the curves beyond the transition point to explore the nature of overstressed and understressed metastable material. It is only within the mean--field model that it is permitted to make the analytic continuation shown in figure 15. The S-shaped curve that results gives a hysteresis loop whose width varies as $(T_c-T)^{\gamma'+\beta'}$. In the mean-field model $\gamma' + \beta' = 3/2$, and so the hysteresis width raised to the power 2/3 would be linear in the temperature. In other models $\gamma' + \beta'$ tends to lie below 3/2.

The remarkable fit of the experimental results of Brereton et al. [20] to the mean-field model is shown in figure 16. The linearity of the solid circles depicting experimental points and the closeness of the slope of the solid line plotted from the parameters of model I seem to indicate that the mean-field model cannot be without predictive power.

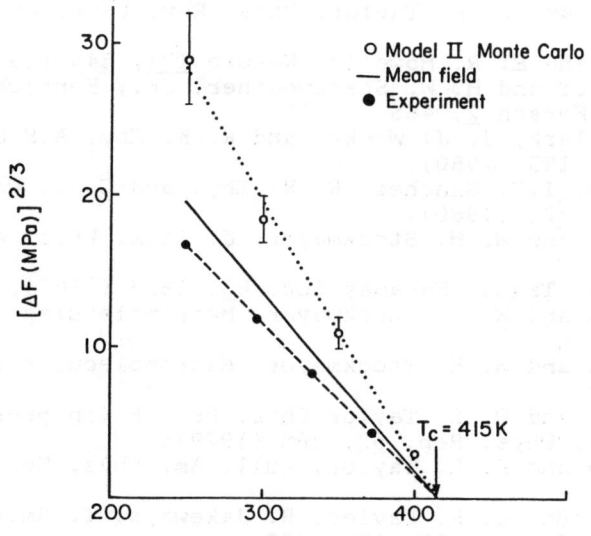

Figure 16. Variation of $(\Delta F)^{2/3}$ with temperature, where ΔF is the width of the hysteresis loop. The solid circles are experimental points from reference 20, while the dashed line is drawn to guide the eye. The solid line refers to model I while the open circles are obtained from computer simulation of model II. The dotted line is a fit to the open circles.

CONCLUSIONS

The richness of the physics to be found in the study of crystalline polymers is matched only by the growing importance of these materials in technological applications. A great deal of research remains to be done before we can say that we have more than a qualitative understanding of the mechanisms involved in phase transitions in these versatile substances.

REFERENCES

1) E. W. Aslaksen, J. Chem. Phys. 57 2358 (1972).
2) R. G. Kepler and R. A. Anderson, J. Appl. Phys. 49, 4490 (1978).
3) S. Duckworth, R. K. Bullough, P. J. Caudrey and J. D. Gibbon, Phys. Lett. 57A, 19 (1976).
4) J. E. McKinney, G. T. Davis, and M. G. Broadhurst, J. Appl. Phys. 51, 1676 (1980).
5) D. Naegele and D. Y.Yoon, Appl. Phys. Lett. 33, 132 (1978).
6) H. Dvey-Aharon, T. J. Sluckin and P. L. Taylor, Phys. Rev. B21, 3700 (1980).
7) T. Furukawa and G. E. Johnson, Appl. Phys. Lett. 38, 1027 (1981).
8) J. D. Clark and P. L. Taylor, Phys. Rev. Lett. 49, 1532 (1982).
9) C. W. Bunn and E. R. Howells, Nature 174, 549 (1954).
10) C. A. Sperati and H. W. Starkweather, Jr., Fortschr. Hochpolym.-Forsch 2, 465 (1961).
11) Edward S. Clark, J. J. Weeks, and R. K. Eby, ACS Symposium Series 141, 183 (1980).
12) J. J. Weeks, I.C. Sanchez, R. K. Eby, and C. I. Posner, Polymer 21, 325 (1980).
13) T. W. Bates and W. H. Stockmayer, J. Chem. Phys. 45, 2321 (1966).
14) T. W. Bates, Trans. Faraday Soc. 63, 1825 (1967).
15) T. W. Bates and W. H. Stockmayer, Macromolecules 1, 12 (1968).
16) T. W. Bates and W. H. Stockmayer, Macromolecules 1, 17 (1968).
17) A. Banerjea and P. L. Taylor Phys. Rev. B (in press).
18) B. Chirikov, Phys. Rep. 52, 265 (1979).
19) V. K. Datye and P. L. Taylor, Bull. Am. Phys. Soc. 29, 408 (1984).
20) M. G. Brereton, G. R. Davies, R. Jakeways, T. Smith, and I. M. Ward, Polymer 19, 17 (1978).

This work was supported by the Materials Research Laboratory program of the N.S.F. under grant DMR81-19425 and by the Army Research Office under grant DAAG29-83K-0168. This paper represents an extended version of a report given at a workshop sponsored by the office of Naval Research at Blacksburg, Virginia.

THE ELASTIC PROPERTIES OF RANDOM NETWORKS

M. F. Thorpe
Physics and Astronomy Department
Michigan State University
East Lansing, Michigan 48824

ABSTRACT

The elastic properties of random networks of springs is richer than the analogous problem with resistors. It is shown that there are two classes of problem. As springs are cut, the network either breaks up at $p^* = p_c$ (class 1) or at $p^* > p_c$ (class 2) where p is the fraction of springs present. Class 1 is ordinary geometrical percolation but class 2 is better viewed as rigidity percolation. An important example of the latter is provided by covalent random networks.

INTRODUCTION

The study of the properties of random elastic systems has been pursued for the past year by a number of groups. A clear picture has yet to emerge and in this brief summary I should like to raise some of the important issues and give a list of references.

It is surprising that this subject is only being pursued now. The random resistor problem has been extensively studied and the reader is referred to the review articles by Kirkpatrick [1] and Straley [2]. In the simplest (class 1) version of this problem, a network of mass points is connected together and described by a potential that couples the (small) motions of the mass points. Sufficient forces must be specified so that when parts of the network are removed, geometrical connection implies elastic connection. In two dimensions this means specifying say, nearest neighbor central forces involving all bonds (ij) and also all angular forces involving the bonds (ij) and (jk). In three dimensions, the dihedral angle forces involving bonds (ij), (jk) and (kl) must be specified in addition. The elastic properties of these networks vanish at percolation p_c, when the network becomes geometrically disconnected. This can be achieved by randomly removing a fraction (1-p) of sites or bonds. It can also be accomplished by cutting holes in a sheet of material like aluminum [3] in the single real experiment that has been done to date.

In class 2 problems, fewer forces are specified and the transition takes place at $p^* > p_c$ which corresponds to <u>rigidity percolation</u>. Much less is known about class 2 problems but there is an important microscopic realization in covalent network glasses. We will briefly discuss these topics in the next few sections.

Class 1 Systems

The underlying geometry is ordinary geometrical percolation but there are two new elastic quantities to monitor in isotropic systems. Simulations have been done in two dimensions on the triangular net [4] and on the honeycomb lattice [5] with bonds randomly removed. The results show that <u>all</u> the elastic constants go to zero at p_c with a universal exponent f

$$c_{ij} \sim (p-p_c)^f$$

with

$$f = 3.5 \pm 0.2 \quad (1)$$

The value of f in 3D is not known. We note that this exponent is quite different from $t \sim 1.3$ for the behavior of the conductivity near p_c in resistor networks [6]. The ratio of the two critical amplitudes v_ℓ (longitudinal and velocity) and v_t (transverse sound velocity) approaches a universal value at p_c [5]

$$v_\ell^2 / v_t^2 = 3.5 \pm 0.2 \quad (2)$$

It is not known whether the equality of (1) and (2) is a coincidence. It is remarkable that 2 is predicted to be universal, that is independent of the geometry of the lattice <u>and</u> of the initial value of v_ℓ^2/v_t^2 when all bonds are present. Continuum effective medium theories [7] show that if ellipses with an aspect ratio of r = b/a (the ratio of major axes) are randomly punched into a 2D continuum then

$$v_\ell^2 / v_\ell^2 = 2[1 + r/(r^2 + 1)] \quad (3)$$

as percolation is approached.

A major unresolved issue is whether square (or cubic) systems become isotropic as percolation is approached. This would imply that the three elastic constants of the square (or cubic) system satisfy the condition that the anisotropy parameter A tends to zero at p_c where

$$A = (c_{11} - c_{12} - 2c_{44})/c_{11} \quad (4)$$

That is the leading order singularities, involving f, cancel in the numerator. By analogy with magnetic systems, we might expect isotropy at p_c, but this has not yet been shown.

Class 2 Systems

Much less is known about these systems. Examples are the triangular net and f.c.c. lattices with central forces only [8]. As bonds are removed, local regions become <u>floppy</u> even though they are geometrically connected [9]. The network loses its elastic properties at $p^* > p_c$ so that it has no elastic properties even though remaining geometrically connected. Constraints arguments show that for central forces

$$p^* = 2z/d \tag{5}$$

Numerical simulations show that (5) is very accurate and could even be exact [10] for reasons that are unknown. Indeed effective medium theory agrees almost exactly with numerical simulations except for a small region around p*. It is not clear whether the small "tail" seen in the simulations is intrinsic or due to finite size effects. The ratio of elastic constants appear to be independent of p (as predicted by effective medium theory).

There is disagreement on the value of f. This is closely tied in with the size of the tail and is hard to estimate from simulations. Feng and Sen [11] find $f = 2.4 \pm 0.4$ whereas Lemieux et al. [12] find that $f = 1.4 \pm 0.2$ (effective medium theory gives $f = 1$). It is certain that class 2 problems are in a different universality class from class 1. The node link model of percolation [13] has been extended to describe class 1 problems and gives $f = 11/3 = 3.67$ in close agreement with (1) It is possible that this model could be extended to class 2 problems in order to clarify the underlying physics.

Covalent Glasses

An important application of class 2 problems is to covalent network glasses like Ge_xSe_{1-x} where $0 < x < 1$. These glasses have covalent bonds (Ge has 4 bonds and Se has 2 bonds) and mean co-ordination $\langle r \rangle = 2x + 2$. Constraints arguments [14,9] on networks with nearest neighbor central and angle forces (but <u>no</u> dihedral angle forces) suggest a phase transition at

$$\langle r \rangle = r_p = 2.4 \tag{6}$$

For $\langle r \rangle < r_p$, the networth is <u>floppy</u>. As $\langle r \rangle$ increases the <u>rigid</u> regions join together and <u>rigidity percolates</u> at r_p. It is, therefore, predicted that there are two kinds of covalent glasses. Those with $\langle r \rangle < r_p$ may be referred to as <u>polymeric glasses</u> as they can be thought of as chains with a few cross links. Those in the $\langle r \rangle > r_p$ may be called amorphous solids.

Recent simulations of the elastic properties [15] confirm these ideas and show that

$$C_{ij} \sim (\langle r \rangle - r_p)^f$$

where $\tag{7}$

$$r_p = 2.4$$

and

   f = 1.5

The experimental implications of these ideas are not yet fully developed. These glasses provide the only microscopic system in which elastic problems of either class 1 or class 2 can be studied.

CONCLUSIONS

As you can see from the number of preprints referenced, this is a rapidly developing area. Many of the issues raised here should become clearer within the next year or so.

REFERENCES

1) Kirkpatrick, S.: Rev. Mod. Phys. 45, 574 (1973).
2) Straley, J. P.: Percolation Structures and Processes in Annals of the Israel Physical Society Vol. 5, Ed. by Deutscher, G., Zallen, R. and Adler, J. (1983)
3) Benguigui, L.: preprint.
4) Feng, S., Sen, P.N., Halperin, B.I. and Lobb, C. J.:preprint.
5) Bergmann, D. J.: preprint.
6) Derrida, B.and Vannimenus, J.: J. Phys. A 15, L557 (1982).
7) Thorpe, M. F. and Sen, P.N.: preprint.
8) Feng, S., Thorpe, M.F. and Garboczi, E.: preprint.
9) Thorpe, M.F.: J. Non Cryst. Solids 57, 355 (1983).
10) Phillips, J.C. and Thorpe, M.F.: preprint
11) Feng, S. and Sen, P. N.: Phys. Rev. Lett 52, 216 (1984).
12) Lemieux, M. A., Breton, P. and Tremblay, A.M.S.: preprint.
13) Kantor, Y. and Webman, I.: Phys. Rev. Lett 52, 1891 (1984).
14) Phillips, J. C.: J. Non Cryst. Solids 34, 153 (1979) and 43, 37 (1981).
15) He, H. and Thorpe, M. F.: preprint.

SOME REMARKS ON DIFFUSION ON FRACTALS

H. Nakanishi
Department of Physics, Purdue University
West Lafayette, Indiana 47907

ABSTRACT

We consider a random diffusion process constrained to take place on a fractal object. For relatively short times, the diffuser travels over length scales for which the underlying object is seen to be self-similar. This is known as the "anomalous diffusion" regime so that the mean square displacement after time t is not proportional to t. We present a first analysis of force-force correlation function in the anomalous regime, and postulate a power law decay for it. This power law is shown to give rise to a remarkable "hyper-universal" relationship between the force-force correlation and the mean square displacement. These results are checked by Monte Carlo simulations of random walks on two- and three-dimensional fractal clusters of two very different nature: percolation and DLA (diffusion limited aggregation). The numerical results are roughly consistent with our hypothesis although their accuracies are not sufficient to test them definitively.

INTRODUCTION

Macroscopic, self-similar objects have been of considerable interest recently from both theoretical [1,2] and experimental [3,4] points of view. They are now recognized to exist in nature rather abundantly in all sorts of contexts, and thus the effects of fractal geometry on various physical phenomena may hold the key to understanding them. An example of an actively pursued area is the fluid flow through porous media, such as the flow of oil through porous rocks [5]; another example is the macroscopic conductivity and dielectric function for inhomogeneous mixtures of good and poor conductors [6]. In many circumstances, the fractal structures themselves are mysterious and objects of intense studies; these include the so-called metal leaves [7] which can be grown at a liquid interface by electrolysis, aggregates of gold colloids in aqueous solution [8], and aggregates of antigens in blood stream formed by the links provided by antibodies [9].

Theoretical interest in fractals is mainly caused by the belief that they somehow embody criticality by their geometrical structures. The connection between geometry and criticality has been known for some time at least for static problems such as the lattice gas model of a liquid-gas transition or the percolation model of a gelation transition. In either case, a droplet model [10] or a more general scaling theory [11] predicts a power law relationship between the characteristic size of a critical "droplet" and the characteristic length scale given by the correlation length. Such a relationship can be interpreted as giving an effective dimensionality to the droplet which is different from the Euclidean dimension and in fact fractional. Thus the concept of fractals fits naturally into the critical phenomena. In fact, the modern theory of renormalization group [12], which entails scaling, is based on the concept of self-similarity, which then leads to fractal structures for fluctuations.

More recently, the study of the so-called diffusion limited aggregation [13] opened the door to the application of the fractal ideas to non-equilibrium problems. This model was first proposed to simulate dust or smoke particle aggregates whose electron micrographs showed fractal structures. Since the shape of the aggregate depends on its entire past history, there does not seem to be an equivalent equilibrium model. Yet, the system clearly exhibits a growing self-similar cluster. Whether this type of irreversible aggregation problem is essentially different from equilibrium critical phenomena or not is an important unsolved problem. Thus, ref. 13 served to open a new area even though later it turned out not to be a good model for dust and smoke aggregates [14].

Thus there are many different kinds of fractals: some are static and associated with critical phenomena, and others are kinetic and formed irreversibly. All of the fractals mentioned above are random fractals in that they are formed in a disorderly fashion. In addition to these, there are regular fractals created according to a regular, deterministic rule. There are no known regular fractals in nature, but there are many known in mathematics; in fact, according to Mandelbrot [1], these are the first fractal structures recognized as such. As we will later present "hyper-universal" nature of anomalous diffusion on fractals, it seems in order to stress the diversity of the so-called fractals.

In the remainder of this section, we will review the definition of fractals and fractal dimensionality and give examples of different kinds of fractals. In the next section, we will review the known aspects of the problem of diffusion on fractals and point out some of the outstanding questions in this area. In the last two sections, we will present the hyper-universal power law relationship between the force-force correlation function for the anomalous diffusion and the mean square displacement, and then show our Monte Carlo results which confirm a power law decay for the correlation function for the first time and which are roughly consistent with the specific hypothesis.

Fractals are highly ramified, fragmented geometrical objects that possess the so-called self-similarity property. This means that the object viewed at different degrees of magnification appear to have essentially the same degree of fragmentation, or in other words, there is no one definite length scale that characterizes the object but there are infinitely many of them. (Strictly, this is possible only for an infinitely large object;

in reality, for finite objects, there is a cut-off length up to which they appear self-similar.) In more formal terms, this amounts to a statement on the two-point concentration correlation function which is defined as follows: take an arbitrary origin on the cluster and calculate the average concentration at distance r from the origin, then average this quantity over all possible positions for the origin. The statement is:

$$C_2(r) \equiv \rho(r) \propto \frac{1}{r^{d-D}} , \qquad (1)$$

where $D<d$ is the fractal dimension. Integrating over a region of radius R, we get

$$N = \int^R d^d r \, \rho(r) \propto R^D , \qquad (2)$$

where N is the average occupied volume in the region of this radius. Equation 2 makes evident the origin of the name fractal "dimension".

Regular fractals can be constructed in many ways, but the simplest and best known kinds are variants of Cantor sets and Koch curves. These can be easily made to have any fractal dimensionality D that is expressible as $\ln q/\ln p$ for rational numbers p and $q<p^d$. We give in figure 1 a variant of the Koch curve and a Sierpinski gasket both of which have $D=\ln 3/\ln 2$.

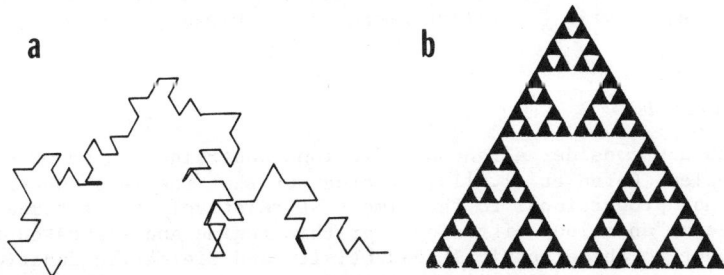

Figure 1. (a) a Koch curve obtained from a line segment by successively replacing the left half of a segment by two segments of lengths equal to the one being replaced; (b) a Sierpinski gasket obtained from a solid triangle by successively eliminating the middle quarter.

Even though both these fractals have the same D, they are obviously very different. One is built "up" from a line (and thus topologically one-dimensional) while the other is built "down" from a two dimensional object. One has no holes and the other is full of them. Thus if there were some physical property that is common to all fractals of a given D (similarly to the "universality" in critical phenomena for given space dimensionality d), it would be quite remarkable.

To complete the introduction to fractals, we give examples of a kinetic (DLA) cluster and a static (percolation) cluster in figure 2. Both of these were constructed on a two-dimensional square lattice, and the fractal dimensionality is about 1.7 for the DLA and 1.9 for the percolation cluster. Once again, despite the similar values of D, the nature of the fragmentation is obviously very different. In addition, percolation gives an entire distribution of cluster sizes at a given probability p while DLA gives a single clus-

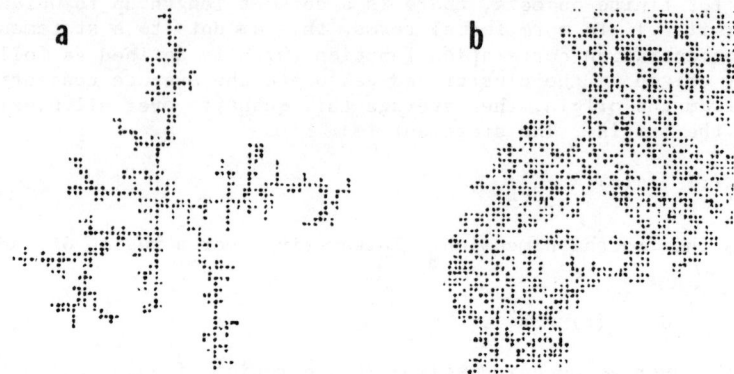

Figure 2. (a) a diffusion limited aggregate; (b) part of a percolation cluster at p_c.

ter that grows without bound. We note that percolation is widely considered to be a model of a gelation transition while the DLA is now believed to be relevant to electrolytic deposition among other things.

Diffusion on Fractals

Let us now consider a random diffusion constrained to occur on a fractal. For this problem, Gefen et al. [15] pointed out that the mean square distance traveled is not proportional to the time t at relatively short times. They coined the term "anomalous diffusion" for this regime and suggested that it may be relevant to the electrical conductivity and dielectric function of a randomly inhomogeneous medium in the high frequency domain. Transport problems are related to diffusion through the Einstein relation linking diffusivity and mobility as is well known. For high frequencies, the motion of charge carriers are restricted to short distances, and thus according to ref. 15 the short time anomalous diffusivity is the relevant quantity.

Thus, while for a normal diffusion,

$$<R(t)^2> \sim Dt , \qquad (3)$$

where D is a constant, for anomalous regime, we have

$$<R(t)^2> \sim D(t)t \sim t^{\frac{2}{2+\theta}} , \qquad (4)$$

for some $\theta>0$. The picture is the following: while the diffuser sees the underlying cluster to be self-similar, diffusion is anomalous with $D(t) \sim t^{-\theta/(2+\theta)}$, and as the distance traveled goes beyond the largest length-scale over which the cluster is self-similar, diffusion crosses over to the normal one with a constant diffusivity D_o. Thus, for example, on an infinite percolation cluster at $p>p_c$, the anomalous regime extends to $t=t_o$ for which $<R(t_o)^2> \sim \xi(p)^2$

where ξ is the connectivity coherence length (see Figure 3). At $p=p_c$, the critical threshold, we will have the anomalous behavior at all (macroscopic) times.

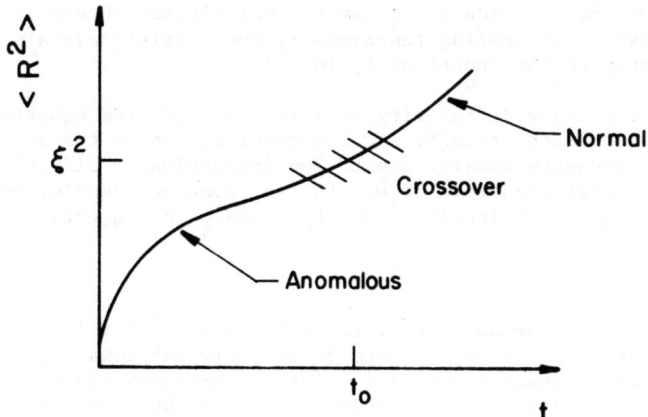

Figure 3. Schematic figure showing the anomalous and normal diffusion regimes.

To understand that such "anomalous" behavior affects the exponent of $<R(t)^2>$ as a function of t and not just the amplitude, it is instructive to consider a simple example of a self-avoiding walk as the underlying cluster. (This can be thought of as diffusion on a long, linear polymer chain.) A self-avoiding walk in three dimensions has a fractal dimension of approximately 5/3. A random walk of N steps on this topologically linear system has the root mean square end-to-end distance proportional to $(N^{1/2})3/5$ for large N. Thus in this case θ is about 4/3. If one regards the track left by a diffusing particle as a cluster, then from equations 2 and 4, its fractal dimension $d_w = 2+\theta$. For percolation in two and three dimensions, the best values for d_w are 2.84 ± 0.05 and 3.68 ± 0.05 respectively [16], to be compared to 2 for the unconstrained random walk. (Incidentally, it is interesting to note that "long-range" diffusion known as Lévy flight has $d_w < 2$. [17])

Gefen et al. [15] gave a scaling description encompassing anomalous and normal diffusion regimes. According to their picture, the mean square end-to-end distance for anomalous diffusion on a percolation cluster of size s is given by

$$<R(t)^2>_s \sim t\xi^{-\theta} f_{\pm}(t/\xi^{2+\theta}, r_s/\xi) , \qquad (5)$$

where $\pm$ indicate $p<p_c$ and $p>p_c$ respectively, and the leading factor is just equal to $D_0 t$. For this to be consistent with equations 3 and 4, we must have $f_-(\infty,\infty)$ = constant and $f_{\pm}(x,\infty) \sim x^{-\theta/(2+\theta)}$ for small x.

In order to average this quantity over all percolation clusters at a given probability p, one must use the cluster size distribution function which itself has a well-known scaling behavior [2]. This renormalizes the exponent of ξ in the leading factor of equation 5. Thus we have

$$\overline{\langle R(t)^2\rangle} \sim t\xi^{-\theta-\beta/\nu} F_{\mp}(t/\xi^{2+\theta}) \;, \tag{6}$$

where $\theta+\beta/\nu$ is known to be equal to μ/ν with μ being the exponent associated with the decrease of the dc conductivity as the percolation threshold is approached from above. The scaling functions $F_{\mp}$ must satisfy certain asymptotic limits similarly to the functions $f_{\mp}$ in (5).

The averaged, effective diffusivity $\overline{D(t)}$ is obtained from equation 6 by differentiation with respect to t, and the contribution to the ac conductivity $\Sigma(\omega)$ (in frequency domain) due to the intra-cluster diffusivity is obtained from the Fourier transform of $\overline{D}$. In the anomalous domain, which obtains if $\xi \to \infty$ or if ω is sufficiently large, this analysis suggests

$$\Sigma(\omega) \sim \omega^{\mu/[\nu(2+\theta)]} \;. \tag{7}$$

The exponent appearing in the above equation has a numerical value of about 0.34 for d=2. However, recent experiments on randomly inhomogeneous gold films [18] yielded an exponent close to 1. This discrepancy has not been resolved satisfactorily, and ref. [18] pointed out many-body effects as a possible cause. In any case, very little work has been done to study the effects of the interactions among many diffusers (but see [19]), and this should be an area to be pursued in the future.

To conclude the review part of this talk, let us now mention an example of what can be considered as a universal result "in the fractal sense" as discussed in the Introduction. Alexander and Orbach [20] pointed out that there is a mapping between the diffusion on a fractal and a lattice vibration problem. The vibrational density of states at low energies goes as $E^{\tilde{D}/2-1}$, where $\tilde{D}$ is termed the "fracton" or spectral dimension. Comparing with the usual result for a full lattice, it is plausible to consider this as a sort of dimensionality which generalizes the Euclidean dimensions of the lattice.

In addition, their mapping gives

$$d_w = 2D/\tilde{D} \;, \tag{8}$$

where D is the fractal dimension of the underlying fractal. Alexander and Orbach [20] made a rather surprising observation that $\tilde{D}$ is numerically very close to 4/3 for percolation clusters in all dimensions, and conjectured that this must be the exact value for all d. Later others [21] extended the conjecture to include other kinds of fractals, such as the DLA. If these conjectures were correct, then equation 8 states that the fractal dimension of a random walk only depends on the fractal dimension of the cluster: thus, universality in the "fractal" sense. Unfortunately, it is known that the conjecture fails for Sierpinski gaskets, percolation backbone [22], and also in disagreement with the ε-expansion at first order in ε for percolation [23]. In fact even for two-dimensional percolation, the mounting numerical evidence [24] suggests that it is not exact. However, the Alexander-Orbach conjecture appears to be an excellent approximation for a number of systems at various d. If one averages over all clusters in the percolation case, though, the average fractal dimension of the random walk will involve the space dimensionality d explicitly and thus ceases to be even approximately universal in the "fractal" sense.

Fluctuations for Anomalous Diffusion

We now turn our attention to the fluctuations associated with anomalous diffusion. Much of what is given in this section is the result of collaboration with Y. Gefen, and also A. Aharony has made important contributions. For an analytic discussion, we will use a Langevin equation formalism [25] of Brownian motion. We start by considering a particle diffusing in a normal Euclidean space. The equation of motion is simply:

$$m \frac{d\vec{v}}{dt} = \vec{f}(t) + \vec{F}(t) , \qquad (9)$$

where m, $\vec{v}$ is the mass and velocity of the particle undergoing Brownian motion and $\vec{f}(t)$ is an external force and $\vec{F}(t)$ is the force describing the interaction of $\vec{r}$ with the many other degrees of freedom of the system. Let also τ^* be the elastic scattering time which gives the scale over which $\vec{F}(t)$ fluctuates. Assuming as usual that there is a slowly varying part to $\vec{F}$ which tends to restore the particle to equilibrium, and assuming this slow part can be approximated by $-\alpha \vec{v}$ where $\vec{v}$ is the slow part of $\vec{v}$, we obtain a Langevin equation:

$$m \frac{d\vec{v}}{dt} = \vec{f}(t) - \alpha v + \vec{F}'(t) , \qquad (10)$$

where $\vec{F}'(t)$ is the rapidly varying part of $\vec{F}(t)$. Thus α is identified as a "friction" constant. Taking the ensemble average of this equation and then solving it for $\vec{f}(t) = 0$, we obtain usual diffusion for long times with the well known result

$$D = \frac{3k_B T}{\alpha} , \qquad (11)$$

linking the diffusion constant D with α. In addition, if, say, a constant external electric field is present and if the particle carries an electric charge e, then the Langevin equation immediately gives a relation for the mobility of the charge:

$$\mu = \frac{e}{\alpha} . \qquad (12)$$

Equations 11 and 12 then lead to the Einstein relation so that diffusion constant is proportional to mobility.

The above derivation of equation 11 assumes that α is constant in t and that the average kinetic energy is given by the equipartition theorem (thus again constant in t) for $d=3$. If the particle is confined to move within a fractal, neither of these assumptions may be valid although the Langevin equation itself must still be valid. One way to get around this difficulty is to assume the Einstein relation to be valid even for the fractal case; then equation 11 immediately follows from equation 12. Instead, we assume: α is a power law in t for large t, or

$$\alpha \sim t^x , \quad x > 0, \qquad (13)$$

and that the average kinetic energy is constant in t. With these assumptions, we can again solve the Langevin equation formally, and study the behavior of $\langle R(t)^2 \rangle$ for large t. The result is that

$$\langle R(t)^2 \rangle \sim t^{1-x} \sim t/\alpha . \qquad (14)$$

Therefore, we regain a relationship of the form of equation 11. Since equation 12 is always valid, we also regain the Einstein relation as expected. For anomalous diffusion, $D(t) \sim t^{-\theta/(2+\theta)}$. Thus we have

$$\alpha \sim t^{\theta/(2+\theta)}, \tag{15}$$

consistent with equation 13. Normal diffusion can be included as $\theta=0$ at this stage.

Our next task is to express α in terms of the force-force correlation function. For normal diffusion in Euclidean space, the solution to this problem is a form of the fluctuation-dissipation theorem [25]:

$$\alpha_o = \frac{1}{2k_B T} \int_{-\infty}^{\infty} \langle \vec{F}(0) \cdot \vec{F}(s) \rangle_o \, ds, \tag{16}$$

where the subscript 0 on α indicates full Euclidean space and $\langle ... \rangle_o$ denotes an equilibrium ensemble average. This relation was obtained under the condition $m/\alpha_o \gg \tau^*$; that is, the heat bath equilibrates much more quickly than the smallest time scale for normal diffusion. Under corresponding conditions, we may translate the steps leading to equation 16 for anomalous diffusion. We must be careful, however, to take account of the possibility that the force-force correlation may not be an exponentially decaying function of time for our problem. Thus finally we obtain a relation analogous to equation 16:

$$\alpha = \frac{1}{k_B T \tau} \int_{-\tau}^{0} \langle \vec{F}(0) \cdot \vec{F}(s) \rangle_o (\tau + s) \, ds, \tag{17}$$

where τ used for the derivation is small on a macroscopic scale but much larger than τ^*. We will additionally assume that this relation can be extended to the time scales appropriate for anomalous diffusion.

From equations 15 and 17, we surmise that the correlation function appearing in equation 17 must indeed have a slowly decaying component. Let us assume that this component can be represented by a power law:

$$\langle \vec{F}(0) \cdot \vec{F}(s) \rangle_o \sim s^{-y} + \text{(short range terms)}, \tag{18}$$

for $s>0$ and symmetric for $s<0$. Once this assumption is made, the integral in equation 17 is simple to carry out, and we obtain terms proportional to τ^{2-y} and τ. The term proportional to τ gives a constant α and leads to normal diffusion, and therefore we must conclude $2-y>1$ and, by comparing with equation 15, $1-y=\theta/(2+\theta)$ for anomalous diffusion, or

$$y = \frac{2}{2+\theta}, \tag{19}$$

satisfying $2-y>1$. Equations 4, 18, and 19 then lead us to a remarkable relation:

$$\langle \vec{F}(0) \cdot \vec{F}(t) \rangle_o \sim \frac{1}{\langle R(t)^2 \rangle} + \text{(s.r.t.)}. \tag{20}$$

This inverse proportionality for the force-force correlation function and the mean-square displacement does not contain any reference to θ, and thus independent of what underlying fractal is used. If we accept this hypothesis, anomalous diffusion on all the fractals reviewed earlier in all spatial dimensions must obey this law.

The above remarkable possibility is limited to anomalous diffusion, and even though θ drops out of the final expression 20, we cannot include the case of normal diffusion (θ=0), because for normal diffusion, there are only short-ranged terms in the correlation function (cf. equation 18). Note that we have made a number of assumptions about α, the correlation function, and time scales, and one or more of these may not be accurate and may invalidate our hypothesis. Also, although the power law decay of the force-force correlation reminds us of the celebrated problem of long-time tails, that problem does not have anomalous diffusion in the sense of equation 4. Thus our analysis appears to have little to do with such a problem. One last remark is that we can obtain a scaling formulation of the force-force correlation function in an analogous manner to that of the mean-square displacement given in [15], and average over, say, clusters in percolation. We shall omit these details in this account.

Numerical Results

In this last section, we report the results of computer simulations of random walks on fractals. The fractals used so far are two- and three-dimensional percolation clusters (at p_c) and DLAs. The principal results are presented in figures 5 and 6, and they show qualitative agreement with the Langevin analysis presented above. However, many quantitative aspects remain ununderstood as described below.

First, we consider $<R(t)^2>$ in figure 4. Figure 4a gives results for percolation clusters generated at p_c on square and simple cubic lattices. In both cases the cluster size is 2500 sites, and the data points are obtained from averaging over 5 million walks of 100 steps each for the square lattice and over 5 million walks of 100 steps and 2.5 million walks of 500 steps for the simple cubic lattice. Thus one cluster for the square lattice and two clusters for the cubic lattice were used for this figure. All random walks used for this and other figures start from (different) randomly selected points on the lattice. The straight diagonal line represents the exact result for a free random walk on a full square lattice, and the points lying on this line are obtained by a very small-scale simulation (10^5 walks of 100 steps). This latter simulation was done to check parts of the algorithm and also to see in general how well this type of simulation reproduces known results. Figure 4b gives analogous results for DLA clusters. For the square lattice, averages are taken over 5 million walks of 100 steps on a cluster of 6000 sites, and for the simple cubic lattice, averages are taken over half a million walks of 100 steps on a cluster of 1706 sites and over 1.5 million walks of 400 steps on a cluster of 4310 sites. The results from random walks on a full square lattice are also included.

As far as $<R(t)^2>$, others have done simulations extending to much longer walks in both two and three dimensions (see for example, [16]) although their sample sizes were far smaller. (Our error bars would be much smaller than the symbols in the figure.) In fact the large sample sizes are not at all needed to study $<R(t)^2>$ alone, but they are essential for the study of the fluctuations as the acceleration-acceleration correlation will be found to decay to zero as a power law of the number of steps. Previous works noted that those long walks (at least a few thousand steps) are necessary to reach asymptotic power laws shown in equation 4. We present this figure principally to emphasize this point: the curvature evident in this figure shows

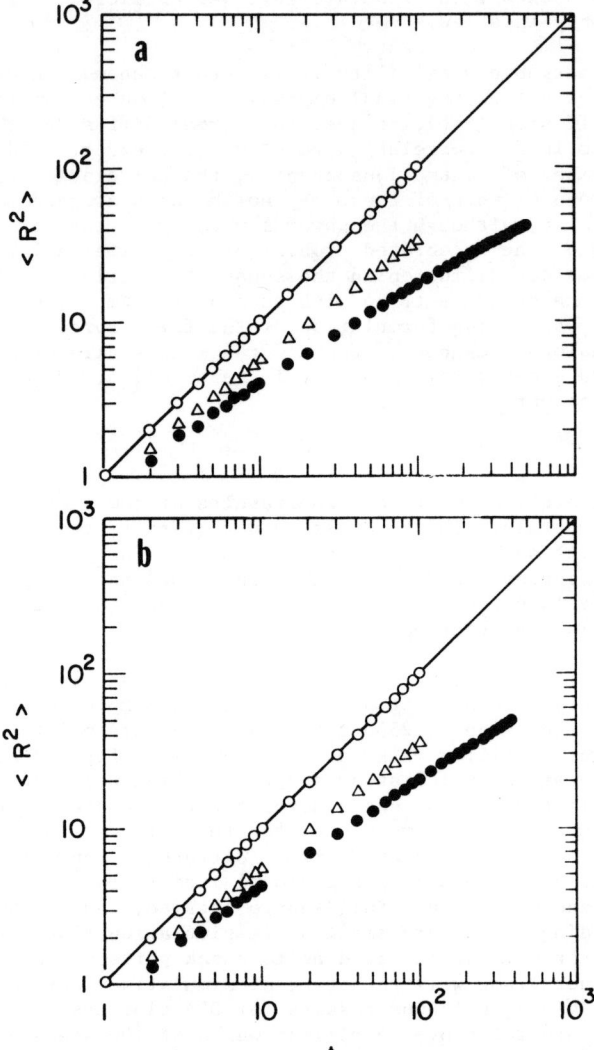

Figure 4. (a) Mean square displacement for random walks on two- and three-dimensional percolation clusters at p_c; (b) Similarly for random walks on two- and three-dimensional diffusion limited aggregates. Results for random walks on a full square lattice is added for reference.

that we have not reached asymptotic, anomalous diffusion regime, and therefore, the prediction of equation 20 is not likely to be quantitatively observed. However, percolation in d=2 and DLA in d=3 appear to be fairly close to asymptopia from this figure, and we will see that those are the cases where closer agreement with equation 20 is obtained.

In figures 5 and 6, we present the magnitude of the acceleration-acceleration correlation (see below about the sign of this quantity) plotted against $<R(t)^2>$ for percolation and DLA, respectively. For a random walk, this correlation function assumes the role of the force-force correlation for a Brownian motion discussed in the previous section. Thus we wish to see a power law of the form of equation 20 in these figures. All the figures are consistent with a power law decay at least qualitatively, and in particular, square lattice percolation and simple cubic lattice DLA exhibit behavior closely matching the exponent of unity (equation 20).

Figure 5a is obtained from two percolation clusters of 2500 sites each and averages are taken over 5 million walks of 100 steps on each cluster. The walks are combined into batches of 1 million each and the standard deviation for the acceleration-acceleration correlation function from the 5 average values for these batches range from 0.1% for the first few steps to 5% or more at the 100th step. Even so, the deviation from cluster to cluster is much larger than that among the 5 batches of walks on the same cluster. This points out a considerable difficulty in obtaining numerical accuracy in this problem. Similarly, figure 5b is for 5 million walks of 100 steps on a percolation cluster of 2500 sites on a simple cubic lattice and for 2.5 million walks of 500 steps also on a (dfferent) cluster of 2300 sites. These walks are also grouped in 5 batches each and the standard deviations are computed from the batch averages. The standard deviation for the acceleration correlation for the 500th step is slightly over 10%. The corresponding errors for the $<R(t)^2>$ are, as mentioned earlier, practically negligible (a bit over 0.1% at the 500th step!).

Figure 6a is for two DLA clusters of size 5910 and 6000 on a square lattice. Averages are taken over 5 million walks of 100 steps on each cluster, by first taking averages for batches of 1 million walks each as before. The error estimates are similar to but slightly larger than those of figure 5a, and again the differences between the results from the two clusters dominate. Figure 6b gives the results for two DLA clusters of size 4310 and 1706 on a simple cubic lattice. Half a million walks on 100 steps on a cluster of 1706 sites and 1.5 million walks of 400 steps on the one of 4310 sites are grouped into 5 batches each and averaged over. The error estimates are comparable to those for figure 5b. In this case the results from two clusters appear to be nearly indistinguishable, and also the curve seems to be a fairly good representation of the relation predicted by equation 20.

Aside from the fact that the decay exponent estimation is not always in good agreement with equation 20, there are a number of surprising and unresolved issues. The potentially most important issue is that we find a rapid and regular oscillation in sign for all acceleration correlation functions as function of t. Namely for all the problems studied, the correlation changes signs at every step, and the absolute values of the correlation function seem to be a smooth function of t decaying with a power law. (This is the reason why the absolute values are plotted in figures 5 and 6.) This contrasts with the case of a random walk on a full hypercubic

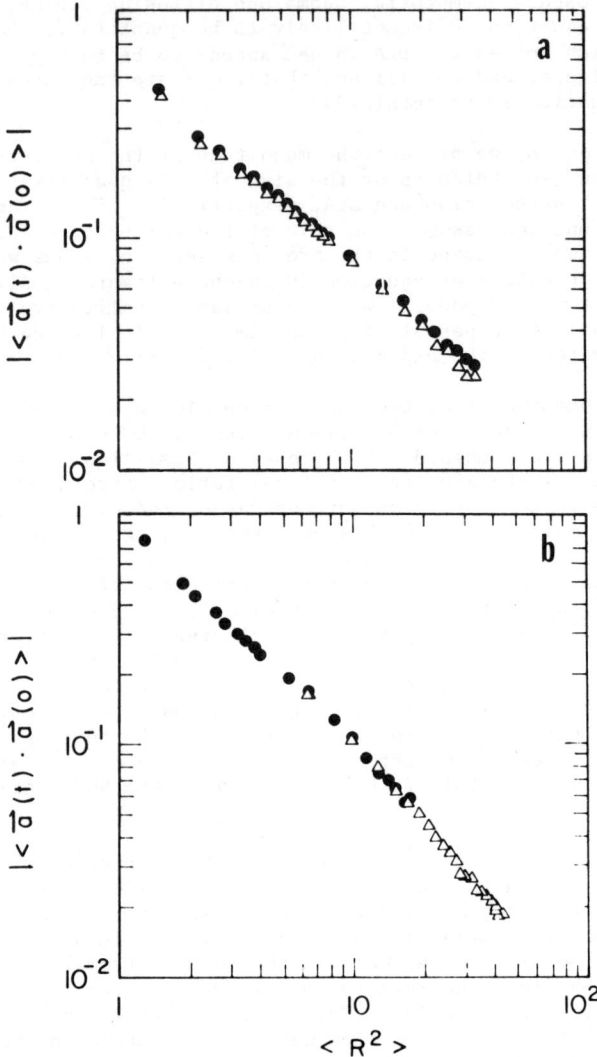

Figure 5. (a) Acceleration-acceleration correlation function plotted against mean square displacement for random walks on percolation clusters at p_c created on a square lattice; (b) Similarly on a simple cubic lattice.

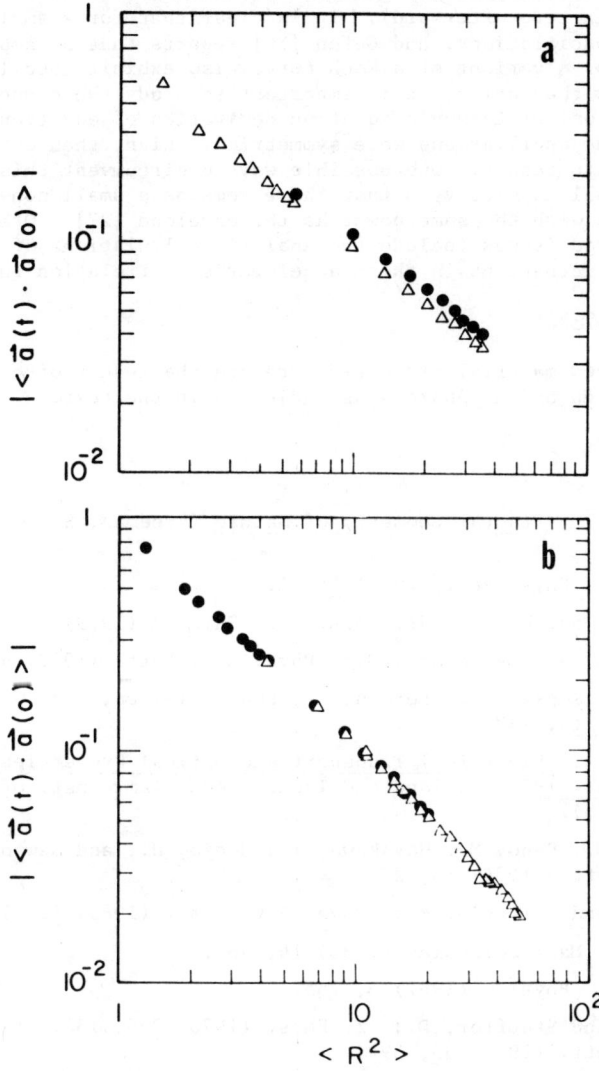

Figure 6. (a) Acceleration-acceleration correlation function plotted against mean square displacement for random walks on diffusion limited aggregates created on a square lattice; (b) Similarly on a simple cubic lattice.

lattice for which the correlation has one oscillation in sign before falling to zero (and remain zero afterward). It is clear that for a small cluster, one expects such oscillations, and Gefen [26] reports that a sample numerical calculation for a variant of a Koch curve also exhibits oscillations. In addition to how they arise, it is important to study the consequences of such oscillations on our Langevin equation derivation of equation 20: if the envelope of the oscillations were symmetric in sign, then our equation 20 would not seem to result. One possible way to circumvent this difficulty is to assume a small asymmetry so that there remains a small non-oscillating component decaying with the same power as the envelope [27]. Other less important unresolved issues include the analytic calculations of equal-time correlation and fluctuations in these acceleration correlation functions.

ACKNOWLEDGEMENT

Much of the new material presented here are the result of collaborations with Y. Gefen and A. Aharony as indicated in the text.

REFERENCES

1) Mandelbrot, B.: <u>Fractal Geometry of Nature</u> (Freeman, San Francisco, 1982).
2) Stauffer, D.: Phys. Rep. (1979) $\underline{54}$, 1.
3) Forrest, S.R. and Witten, Jr., T.A.: J. Phys. A (1979) $\underline{12}$, L109.
4) Kapitulnik, A. and Deutscher, G.: Phys. Rev. Lett. (1982) $\underline{49}$, 1444.
5) Chandler, R., Kopkik, J., Lerman, K., and Willemsen, J.: J. Fluid Mech. (1982) $\underline{119}$, 249.
6) Straley, J.P. in <u>Electrical Transport and Optical Properties of Inhomogeneous Media - 1977</u>, Garland and Tanner, eds. (Am. Inst. of Physics, New York, 1978).
7) Matsushita, M., Sano, M., Hayakawa, Y., Honjo, H., and Sawada, Y.: Phys. Rev. Lett. (1984) $\underline{53}$, 286.
8) Weitz, D.A. and Oliveria, M.: Phys. Rev. Lett. (1983) $\underline{52}$, 1433.
9) Benedek, G.: Macromolecules (1983) $\underline{16}$, 434.
10) Fisher, M.E.: Physics (1967) $\underline{3}$, 255.
11) Kiang, C.S. and Stauffer, D.: Z. Phys. (1970) $\underline{235}$, 130: Stauffer, D.: Phys. Rev. Lett. (1975) $\underline{35}$, 394.
12) Wilson, K.G. and Kogut, J.: Phys. Rep. (1974) $\underline{12}$, 75.
13) Witten, Jr., T.A. and Sander, L.M.: Phys. Rev. Lett. (1981) $\underline{47}$, 1400; Phys. Rev. B (1983) $\underline{27}$, 5686.
14) Meakin, P.: Phys. Rev. Lett. (1983) $\underline{51}$, 1119; Kolb, M., Botet, R., and Jullien, R.: Phys. Rev. Lett. (1983) $\underline{51}$, 1123.
15) Gefen, Y., Aharony, A., and Alexander, S.: Phys. Rev. Lett. (1983) $\underline{50}$, 77.
16) Havlin, S. and Ben-Avraham, D.: preprint.
17) Hughes, B.D., Schlesinger, M.F., and Montroll, E.W.: Proc. Natl. Acad. Sci. (1981) $\underline{78}$, 3287; see also Halley, J.W. and Nakanishi, H. (1984), preprint.

18) Laibowitz, R.B. and Gefen, Y.: Phys. Rev. Lett. (1984) $\underline{53}$, 380.
19) Gefen, Y. and Halley, J.W.: preprint.
20) Alexander, S. and Orbach, R.: J. Phys. (Paris) (1982) $\underline{43}$, L625.
21) Leyvraz, F. and Stanley, H.E.: preprint; Meakin, P. and Stanley, H.E.: preprint.
22) Stanley, H.E. and Coniglio, A.: Phys. Rev. B (1984) $\underline{29}$, 522.
23) Harris, A.B., Kim, S., and Lubensky, T.C.: Phys. Rev. Lett. (1984) $\underline{53}$, 743.
24) Hong, D.C., Havlin, S., Herrmann, J., and Stanley, H.E.: preprint.
25) For an elementary account see, e.g., Reif, F.: <u>Fundamentals of Statistical and Thermal Physics</u> (McGraw-Hill, New York, 1965).
26) Gefen, Y.: private communication.
27) Fisher, M.E.: private communication.

NATURE OF ELECTRONIC WAVE FUNCTIONS IN DISORDERED SYSTEMS

C. M. Soukoulis
Ames Laboratory and Department of Physics
Iowa State University, Ames, Iowa 50011

E. N. Economou
Department of Physics, University of Crete and Research Center of Crete,
Heraklio, Crete, Greece

ABSTRACT

The question of how to quantitatively characterize the wave functions in disordered systems is examined. We discuss the following relevant quantities: the phase coherence length ℓ, the localization length λ, the amplitude fluctuation length ξ, the participation ratio p and the fractal dimensionality D. Various techniques for calculating these quantities are mentioned and relevant results are presented.

INTRODUCTION

The electronic eigenfunctions in disordered systems are complicated objects. Recent explicit calculations by Soukoulis and Economou [1] show that the eigenfunctions have strong amplitude fluctuations of various spatial extents. This is clearly seen in figure 1 where we plot the probability density $|c_n|^2$ for a 2-d (squared) tight binding model of a disordered system with diagonal disorder of total width W. Even for weak disorder (W≈1) the wave function has strong fluctuations and does not occupy the whole available space. For the 2-d case we have independent evidence that all eigenstates are exponentially localized but for weak disorder (W=1 for E=0) we expect the localization length to be $\lambda \sim 10^4$. Therefore the case shown in figure 1a is like any extended state but it has all these strong fluctuations. As disorder increases (figures 2b-2d) the wave function becomes localized within the size of the system studied. Of course, even for these strongly localized eigenstates there are strong fluctuations up to a length which is roughly equal to λ.

As a result of these fluctuations only a fraction of the available space is effectively utilized by the eigenfunctions.

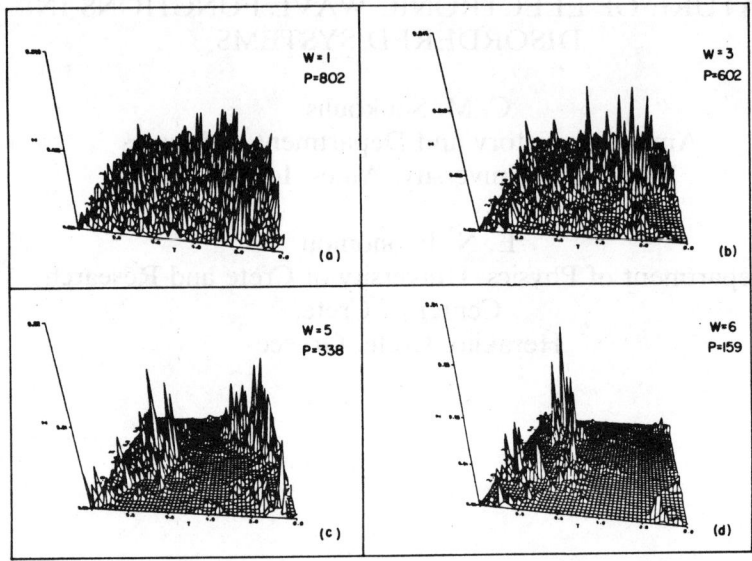

Fig. 1. Plot of the probability of finding a particle at site n, $|c_n|^2$, for a 50×50 square tight binding model with diagonal disorder of total width W. The energy $E \simeq 0.45$ and P is the participation ratio.

In contrast to these findings, <u>ordinary transport theory</u> is based upon the assumption that the amplitude of the eigenfunctions is essentially unaffected by the disorder, while the phase is randomized within a characteristic phase coherence length, the so called mean free path ℓ. The latter can be defined by the relation

$$\langle G(\underline{m},\underline{n}) \rangle = G_0(\underline{m},\underline{n}) \exp\left[-\frac{|\underline{m}-\underline{n}|}{2\ell}\right], \qquad (1)$$

where $G_0(\underline{m},\underline{n})$ is the off-diagonal matrix element of the periodic Green's function between the points $\underline{m}$ and $\underline{n}$, and $\langle G(\underline{m},\underline{n}) \rangle$ is the average value of the same matrix element for the disordered system. For weak disorder, ℓ is given by

$$\ell = |\underline{v}|\tau \qquad (2)$$

where $|\underline{v}|$ is the magnitude of the velocity $\underline{v} = \partial E(\underline{k})/\hbar\partial\underline{k}$ and τ is the relaxation time. In the weak scattering limit, where the assumption of a uniform amplitude is supposed to be valid, the mean free path ℓ is the quantity which controls the dc conductivity σ_0 [2,3]:

$$\sigma_0 = \frac{2}{(2\pi)^d d} \frac{e^2}{\hbar} S_0 \ell . \qquad (3)$$

In equation 3, d is the dimensionality of the space, and S_0 is the area of the Fermi surface (for d = 3), or the length of the Fermi line (for d = 2), or $S_0 = 2$ (for d = 1).

As the disorder increases, the amplitude of the eigenfunctions ceases to be uniform. As a result, the dc conductivity σ does not coincide anymore with the semiclassical expression σ_0 given by equation 3. One expects intuitively that $\sigma < \sigma_0$, and that the difference $\sigma_0 - \sigma$ would increase as the amplitude fluctuations become larger (in size and extent).

Characterization of the Amplitude

Localization length λ

It seems well established now that disorder may lead, if strong enough, to eigenfunctions whose amplitide decays to zero for large distances. Although there is no rigorous proof (except in 1-d), it is usually assumed that the decay is exponential on the average. The characteristic length λ, which determines this exponential decay is called the localization length, and is defined by the relation

$$\langle |\psi(r)| \rangle_g \sim \exp\left[-\frac{r}{\lambda}\right], \text{ as } r \to \infty \tag{4}$$

where the symbol $\langle\rangle_g$ indicates the geometric mean. The main effect of this disorder induced localization is to make the T=0, dc conductivity σ to approach zero as the linear dimension of the specimen L approaches infinity

$$\langle \sigma(L) \rangle_g \sim \exp\left[-\frac{2L}{\lambda}\right], \text{ as } L \to \infty. \tag{5}$$

It has been convincingly demonstrated that in 1-d disordered systems all eigenstates (except some special pathological cases [4]) are exponentially localized no matter how weak the disorder is [2,3]. It is widely believed that the same is true for 2-d disordered systems, although proposals for a power law localization have been advanced. On the other hand, for 3-d disordered systems the prevailing belief is that for not so strong disorder the spectrum is separated by critical energies termed mobility edges into alternating regions of extended ($\equiv$ non decaying) and localized eigenstates. As the disorder increases the regions of extended states may disappear altogether and the whole spectrum may consist of localized eigenstates.

Fluctuation length ξ

For extended states in 3-d disordered systems a length ξ has been introduced characterizing the spatial extent of the largest (in size) fluctuation. If the eigenfunction is averaged over length scales equal to or larger than ξ, it would look uniform. Obviously fluctuations are characterized not only by their extent but by their magnitude as well. The latter can possibly be defined as the ratio of an appropriately averaged maximum value of $|\psi(r)|$ over an appropriately averaged minimum value of $|\psi(r)|$. Very little attention has been given to the question of the magnitude and its possible correlation with ξ. Preliminary unpublished work by the authors of the present article based on the potential well analogy [5] indicates that the magnitude of the largest fluctuation equals ξ/a' where a' is comparable to the interatomic distance. This simple result leads to a reduction of the conductivity according to the formula

$$\sigma = \sigma_0 \frac{a'}{\xi}. \tag{6}$$

The ξ dependence of equation 6 coincides with the predictions of the scaling theory and the field theories (for a brief review see ref. [3]).

Since localized states exhibit considerable fluctuations before eventually their amplitude becomes negligible, a legitimate question to be raised is how to characterize these fluctuations. Although this question has not been examined seriously, it is usually assumed that for localized states the role of ξ is played by λ. This assumption deserves more attention especially in the case of 2-d weakly disordered systems where λ can become very large.

Participation Ratio p

The participation ratio characterizes the fraction of space effectively occupied by an extended eigenfunction, i.e., $p = N_{eff}/N$; N_{eff} is the number of atomic sites where $\psi(r)$ is appreciable and N is the total number of sites. More precisely p is defined by

$$p^{-1} = N \sum_n |c_n|^4 \tag{7}$$

where $|c_n|^2$ is the probability of finding the particle at the site n. The participation ratio appears in phonon-mediated self energies and interactions, which in the static limit lead to an interaction part in the Hamiltonian of the form

$$H_{int} \sim \int |\psi(\underline{r})|^4 d\underline{r} \sim \frac{1}{Np} . \tag{8}$$

It is worthwhile to point out that preliminary results based on the potential well analogy [5] give for extended states that

$$p = \frac{a'}{\xi} \tag{9}$$

which combined with equation 8 lead to an enhancement of the lattice mediated interaction by a factor of ξ/a'. Equation 9 is consistent with numerical results for p.

Fractal Dimensionality D

The quantity D can be defined if the integral of the probability density $|\psi(r)|^2$ within a sphere of radius L is proportional to L^D with D independent of L. For a disordered eigenfunction the result depends strongly on where the center of the sphere is placed. To avoid this difficulty a weighted average over all positions of the center is taken; the weight is the probability density of finding the particle at each point. Thus the fractal dimensionality is defined as the L independent exponent in the relation

$$A(L) = \text{const } L^D \tag{10}$$

where A(L) is the density correlation function

$$A(L) = \int d\underline{r} \, |\psi(\underline{r})|^2 \int_0^L d\underline{r}' \, |\psi(\underline{r}'+\underline{r})|^2 . \tag{11}$$

For uniform extended states the fractal dimensionality coincides with the Euclidean dimensionality: D=d. Thus for extended states and $L > \xi$, D=d; as a result a non-trivial fractal dimensionality can be defined only in the range $a' \ll L \ll \xi$. For localized states a fractal dimensionality can only be defined for lengths less than the effective extent of the eigenfunction; beyond this length, A(L) saturates approaching asymptotically one.

Recently Soukoulis and Economou [1] have calculated numerically the density correlation function A(L) defined by equation 11 in order to check whether a fractal dimensionality could be defined for an eigenfunction in a disordered system. Their numerical results strongly suggests that D is well defined for length scales a'<<L<<λ, ξ and that D is a continuous, decreasing function of the disorder. The most interesting case is for d=3 at the mobility edge, where both λ and ξ are infinite. The fractal dimensionality at the mobility edge was estimated [1] to be 1.7±0.3.

It is very interesting to check experimentally the fractal character of wave functions in disordered systems. A possible probe might be the frequency dependent conductivity $\sigma(\omega)$. In a 3-d disordered system where a mobility edge exists for low frequencies (long times) we are going to see a regular behavior of $\sigma(\omega)$ which is characteristic of uniform extended states. For high frequencies (short times) we are going to probe the local behavior of the states which are fractal like and therefore we will obtain a different frequency dependent of $\sigma(\omega)$. I want to emphasize that the ideas are very speculative in nature.

The quantity D in addition to its ability to characterize quantitatively the shape of the amplitude fluctuations may prove extremely important physically if it turns out that it can determine critical exponents for disordered systems, as the ordinary dimensionality does for ordered systems.

RESULTS

Quantities like the average density of states, the mean free path ℓ, the quasi-free carrier conductivity σ_0, etc., can be calculated rather successfully by mean field theories most notably by the so called Coherent Potential Approximation (CPA). The CPA has been employed both for simple model systems such as tight-binding Hamiltonians and for realistic systems such as amorphous semiconductors.

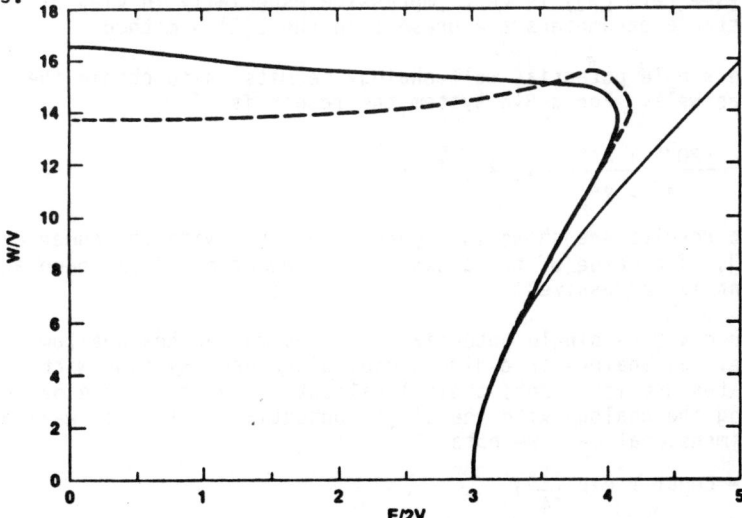

Fig. 2. Mobility edge trajectory for a simple cubic tight binding model with a diagonal disorder of rectangular shape and total width W. V is the off-diagonal nearest neighbor matrix element. The heavy solid line is based on equation 14 while the dashed line is based on the L(E)-method. The solid thin line is the CPA band edge.

Recently Economou et al. [5] have shown that the problem of localization in a disordered system can be approximately mapped into that of a bound level in a shallow single potential well. The extent a of the equivalent well has been taken as proportional to the mean free path ℓ

$$a \propto \ell \qquad (12)$$

and the depth V_0 of the equivalent well is proportional to $1/\sigma_0 a^d$

$$V_0 \propto \frac{1}{\sigma_0 a^d} \sim \frac{1}{S\ell^{d+1}} . \qquad (13)$$

For d=3, a bound level in a well appears only when the product $V_0 a^2$ exceeds a critical value. By analogy, in a 3-d disordered system, localized states appear only when the dimensionless quantity $S\ell^2$ is less than a critical value, which can be obtained by fitting the numerical value at the center of the band of a simple cubic tight binding model [6,7]. The final result is that the mobility edge in a 3-d disordered system is obtained from the simple relation

$$S\ell^2 \approx 9 . \qquad (14)$$

It is worthwhile to point out that for an energy E well inside a band, S is proportional to a'^{-2} so that equation 14 gives $\ell \sim a'$ as the localization criterion in agreement with Mott's [2] proposal. On the other hand, for weak disorder, the mobility is close to the band edge and S there is much smaller than a'^{-2}. This means that for weak disorder the eigenstates can become localized while their mean free paths are considerably larger than the interatomic distance.

In figure 2 we show explicit results based on equation 14 for the trajectory of the mobility edge. The agreement with the results based on the L(E) - method (see ref. [3]) is very impressive especially in view of the fact that no adjustable parameters are present in the L(E) - method.

The single potential well analogy permits us to obtain the localization length as well. For a 3-d system the result is

$$\lambda \approx \frac{(20/S + A\ell^2)\ell}{\ell_c^2 - \ell^2} ; \quad \ell < \ell_c . \qquad (15)$$

Explicit results are shown in figure 3 together with the numerical data of ref. [5]. The value of the constant A in equation 15 has been adjusted. The agreement is impressive.

For $d < 2$, a single potential well, no matter how shallow, always binds a particle. By analogy in d-dimensional disordered systems with $d < 2$ all eigenstates are localized; their localization lengths λ can be obtained by employing the analogy with the single potential well. For weak disorder in the 2-dimensional case we obtain

$$\lambda \approx \text{const } \ell \exp\left[\frac{S\ell}{4}\right] \qquad (16)$$

where the constant, for E at the center of the band, is 2.72, and $S = 4\sqrt{2}\pi$ (in units of inverse lattice spacing). Explicit results are shown in figure 4 together with the numerical data of refs. [6] and [7]. Given that there is only one adjustable parameter in the theory the agreement is very good except for very weak disorder where the numerical data are systematically below the theoretical results.

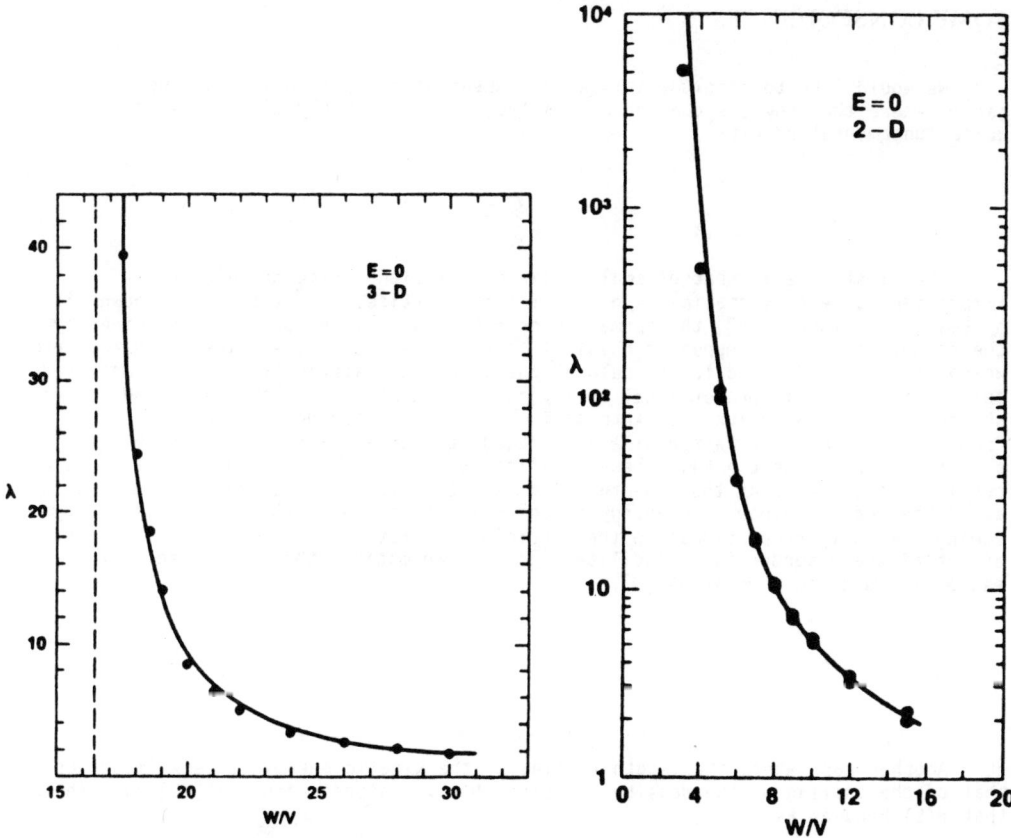

Fig. 3. Localization length λ (in units of lattice spacing) for the center of the band of a simple cubic tight binding model with diagonal disorder of total width W. V is the off-diagonal matrix element. The solid line is the result based on the single potential well analogy and the dots are numerical data of ref. [6].

Fig. 4. Localization length λ (in units of lattice spacing) for the center of the band of a square tight binding model with diagonal disorder of total width W. V is the off-diagonal matrix element. The solid line is the result based on the single potential well analogy and the dots are numerical data of refs. [6,7].

In the 1-d case the final result based on the single potential well analogy is

$$\lambda = 2\ell . \qquad (17)$$

One may argue that ℓ in equation 17 must be obtained from the geometric mean of G (m,n) and not the arithmetic, because the former and not the latter is representative of the ensemble. With this interpretation of ℓ equation 17 becomes exact.

SOME CONSEQUENCES

We would like to conclude this brief presentation by pointing out two cases, where the previous considerations lead in an elegantly simple way to quite fundamental results.

The first case is that of small polaron formation, where the electron immobilizes itself by statically distorting the lattice. As it has been shown by Emin and Holstein [8], the formation or not of a small polaron is determined by the competition of two opposing physical factors. A "repulsive" kinetic energy term which scales as $1/L^2$ and favors delocalization and an "attractive" lattice mediated self-energy which is proportional to $\int |\psi|^4 dr$ and hence it scales as $1/L^3$ and favors shrinkage of the wave function into atomic size. The disorder reduces the "repulsive" term (by a factor of $a'/\min(L,\xi)$) and enhances the "attractive" term (by a factor of ξ/a' for $L > \xi$). These modifications can be understood physically as a result of the fact that the eigenfunction is not uniform, and it does not utilize all available space. Hence, the energy to compress it is lower, while the self-energy (being inversely proportional to the participation ratio) is higher. Thus the net effect of the disorder is to facilitate polaron formation; the larger the ξ, the easier it is to form polarons [9].

Another case where the length scaling of the kinetic energy is very important is that of the tailing of the density of states $N(E)$. Halperin and Lax [10] have shown that $N(E)$ behaves as

$$N(E) \propto \exp\left[-\left|\frac{E}{E_0}\right|^{2-\frac{d}{2}}\right] . \tag{18}$$

The factor 2 that divides the dimensionality d in the exponent is the same as the exponent in the length scaling of the kinetic energy: $1/L^2$. In three and higher dimensionality and for $L < \xi$ the kinetic energy scales [11] as $1/L^d$. Hence the exponent $|E/E_0|$ in equation 18 must become $2-(d/d)=1$ and the DOS must be given by

$$N(E) \propto \exp\left[-\left|\frac{E}{E_0}\right|\right] \; ; \; d > 2 . \tag{19}$$

This naturally resulting exponential tail in the DOS may provide a convincing interpretation of the exponential absorption edge appearing almost universally in disordered systems.

On the other hand, one can argue [12] that in 1-d we are almost always in the regime, where the kinetic energy scales as $1/L^2$ so that

$$N(E) \propto \exp\left[-\left|\frac{E}{E_0}\right|^{3/2}\right] \; ; \; d = 1 \tag{20}$$

which is the exact result in 1-d.

REFERENCES

1) Soukoulis, C. M. and Economou, E. N.: Phys. Rev. Lett. (1984) $\underline{52}$, 565 and unpublished.

2) Mott, N. F. and Davis, E. A.: Electronic Processes in Non-Crystalline Materials, 2nd Ed. (Clarendon, Oxford, 1979).

3) Economou, E. N.: Green's Functions in Quantum Physics, 2nd Ed. (Springer, Heidelberg, 1983).

4) Soukoulis, C. M. and Economou, E. N.: Phys. Rev. (1981) B$\underline{24}$, 5698.

5) Economou, E. N. and Soukoulis, C. M.: Phys. Rev. (1983) B$\underline{28}$, 1093 and Economou, E. N., Soukoulis, C. M. and Zdetsis, A. D.: Phys. Rev (1984) B$\underline{30}$, 1686.

6) MacKinnon, A. and Kramer, B.: Z. Phys. (1983) B$\underline{53}$, 1.

7) MacKinnon, A. and Kramer, B.; Phys. Rev. Lett. (1982) $\underline{49}$, 695.

8) Emin, D. and Holstein, T.: Phys. Rev. Lett. (1976) $\underline{36}$, 323.

9) Cohen, M. H., Economou, E. N. and Soukoulis, C. M.: Phys. Rev. Lett. (1983) $\underline{51}$, 1202.

10) Halperin, B. I. and Lax, M.: Phys. Rev. (1966) B$\underline{14}$, 722 and Phys. Rev. (1967) $\underline{153}$, 802.

11) One argument to understand the different length scaling of the kinetic energy is the following: For a random walker in a regular periodic system after time t the square of the average displacement $<r^2>$ from its starting point goes proportional to t. Therefore by an uncertainty principle argument $((\Delta E)(\Delta t) \sim h)$ one obtains that energy scales as $1/L^2$ in any dimension. Now if the random walker moves in a fractal space or in a highly non uniform space after time t, $<r^2> \sim t^x$ where $x < 1$. This is so, because the random walker spends most of its time wandering around its starting point. Using $(\Delta E)(\Delta t) \sim h$ one obtains that energy scales as $1/L^{2/x}$, $x < 1$. There is a lot of numerical evidence that $x = 2/d$ for disordered systems.

12) Soukoulis, C. M., Cohen, M. H. and Economou, E. N.: Phys. Rev. Lett. (1984) $\underline{53}$, 616.

REFERENCES

DEFECT SIMULATION AND SUPERCOMPUTERS

A. B. Kunz
Department of Physics
Michigan Technological University, Houghton, MI 49931

ABSTRACT

In this report we look briefly at one approach to the study of point defects in solids and to the profitable adaptations one might make to programming structures if one is to achieve maximal efficiency in coding such methods on a variety of scalar, vector and matrix oriented computers. Most common mainframe and supermini computers fall into the scalar category, whereas most common array processors or super computers are vector type processors. The recently announced FPS164/MAX, the Denelcore HEP and several others require multiprocessing for greatest efficiency, and are loosely termed matrix oriented computers. Finally, an example of core excitons in alkali-halides will be included to illustrate the potential of these techniques.

INTRODUCTION

The theory of point defects in crystalline solids has had a long and honorable history and received an early exposition in the book of Mott and Gurney [1]. The field has expanded and received numerous more recent expositions [2,3]. It is not the purpose of this report to survey the field, but rather to consider one approach alone, and the ramifications of adapting this methodology to several classes of computer. These considerations apply equally well to many alternate approaches to defect simulation as well as to many other problems. The methodology employed here is based upon the Hartree-Fock technique, augmented by direct application of Many Body Perturbation Theory (MBPT). These methods are briefly described in the next section. Some computational considerations of these methods are discussed in the third section and some simple results are shown in the fourth section. Finally, in the last section a brief description of several new applications is given; these applications being made possible only by the advent of supercomputers.

Methodology

In all our studies the Unrestricted Hartree Fock (UHF) method is employed as a starting point, as is the normal non-relativistic Hamiltonian:

$$H = \sum_{i=1}^{n} -\frac{h^2}{2m} \nabla_i^2 - \sum_{i=1}^{n} \sum_{I=1}^{n} \frac{e^2 Z_I}{|\vec{r}_i - \vec{R}_I|}$$

$$+ \frac{1}{2} \sum_{i,j=1}^{n,} \frac{e^2}{|\vec{r}_i - \vec{r}_j|} + \frac{1}{2} \sum_{I,J=1}^{N} \frac{Z_I Z_J e^2}{|\vec{R}_I - \vec{R}_J|} .$$

(1)

In this report lower case letters designate electronic properties and upper case letters refer to nuclear properties. The ith electron has coordinate, $\vec{r}_i$ ($\vec{x}_i$ including spin), mass, m, charge, e. The Ith nucleus has atomic number, Z_I, position, $\vec{R}_I$. It is assumed the nuclei are infinitely massive and the Born-Oppenheimer approximation is used. Ideally, one would like to solve the n-electron Schrodinger equation:

$$H \psi_\alpha(\vec{x}_1 \text{ -- } \vec{x}_n) = E_\alpha \psi_\alpha (\vec{x}_1 \text{---} \vec{x}_n).$$

(2)

However, exact solutions to (2) are seen as impractical for these systems, and we resort to the UHF approximation. One assumes:

$$\psi_\alpha(\vec{x}_1 \text{---} \vec{x}_n) \simeq (n!)^{-1/2} \det | \phi_i^\alpha (\vec{x}_i) |$$

(3)

That is, one approximates the solution by a single Slater determinant of one electron orbitals, ϕ_i^α. In the UHF approximation, these orbitals are constrained to form an orthonormal set and to be eigenstates of the z component of spin. The orbitals are not constrained to be double occupied or to have well defined symmetry properties. Choosing the orbitals variationally yields the Hartree-Fock equation:

$$F(\rho^\alpha) \; \phi_i^\alpha = \epsilon_i^\alpha \; \phi_i^a,$$

(4)

where

$$\rho^\alpha(\vec{x},\vec{x}\cdot:) = \sum_{i=1}^{n} \phi_i^\alpha (\vec{x}) \; \phi_i^\alpha(\vec{x}'),$$

(5)

and

$$F(\rho^\alpha) = -\frac{\hbar^2}{2m}\nabla^2 - \sum_I \frac{e^2 Z_I}{|\vec{r} - \vec{R}_I|} + e^2 \int \frac{\rho^\alpha(\vec{x}\cdot,\vec{x}\cdot)}{|\vec{r} - \vec{r}\cdot|}$$

$$- e^2 \rho^a(\vec{x},\vec{x}\cdot)/|\vec{r} - \vec{r}\cdot| \; P(\vec{x}\cdot\vec{x}). \tag{6}$$

$P(x\cdot,x)$ is the operator which replaces coordinate x with $x\cdot$.

For a solid system with low symmetry, such as a solid with a point defect, solutions to even the UHF system of equations are difficult to encompass. It is useful to make use of the arbitrariness of the Fock equation and to rotate to local solutions if possible. The way this is done has been given by Kunz and Klein [4] and further developed by Kunz [5]. This technique is most useful for non-metallic systems such as are studied here. One formally partitions the system into two parts, the cluster to be studied and its environment, which is assumed known. In our case the environment is treated as being the same as in a perfect solid case. The environment then provides a known external potential for the cluster. The cluster is solved self-consistently in the field of the environment. This is fully discussed in the references cited above. Currently, efforts involving several groups are in motion to permit the environment to be treated self-consistently as well. This is discussed briefly in the final section.

The UHF method omits correlation effects. A brief description of the methods being currently employed is in order. Correlation methods to be used for extended systems are constrained by size consistency considerations [6,7]. Our group has chosen to use those based upon multi-reference many body perturbation theory (MR-MBPT). Let the exact Hamiltonian be partitioned into a "simple Hamiltonian," H_o, chosen to be the sum of the one-body Fock operators for the n-body system. Thus

$$H = H_o + V \tag{7}$$

and

$$H_o \phi_i = w_i \phi_i.$$

Consider the first n eigenstate of H_o separately. There may be no state of H_o degenerate with these n-states unless it is also included in the n. P is a projector onto the space of these n states and is,

$$P = \sum_{i=1}^{n} |\phi_i\rangle\langle\phi_i|.$$

Consider,
$$H\Psi = E\Psi = (H_o + V)\Psi$$
assume we wish to find state Ψ, say. Then
$$H_o\Psi = (E - V)\Psi \text{ and}$$

$$(1 - P)(H_o - w_1)\Psi = (1-P)(E - w_1 - V)\Psi.$$

Commuting $(1 - P)$ with $(H_o - w_1)$ permits one to obtain formal solution for Ψ

$$P\Psi = \Psi - (H_o - w_1)^{-1}(1 - P)(E - w_1 - V)\Psi.$$

Now
$$P\Psi = \sum_{j=1}^{n} \phi_j \langle \phi_j | \Psi \rangle = \sum_{j=1}^{n} \pi_j \phi_j = \Phi.$$

Φ is of course unknown. Nevertheless one finds

$$\Psi = T\Phi,$$

(8)

where

$$T = \{1 - (H_o - w_1)^{-1}(1-P)(E - w_1 - V)\}^{-1}.$$

(9)

One may obtain the energy from the secular equation:

$$(E - w_i)\pi_i = \sum_{k=1}^{n} \pi_k \bar{V}_{ik},$$

(10)

where

$$V_{ij} = \langle \phi_i | VT | \phi_j \rangle$$

(11)

From a utilitarian point of view these equations, (8)–(11), are not final in that the unknown energy E occurs in the denominator of Eq. (9)., and results in size consistency problems unless treated properly. This difficulty may be circumvented here, as in Rayleigh-Schrodinger perturbation theory, by using the first order approximation to the energy. To do this, and to solve these equations one expands the inverse in Eq. (9) in a power series

$$T = 1 + (H_o - W_1)^{-1} (1 - P) (E - W_1 - V)]^n + - - - - - -$$

$$+ [(H_o - W_1)^{-1} (1 - P) (E - W_1 - V)]^n + - \quad (12)$$

In this case the first approximation to E is found by solving

$$(E - W_i)\pi_i = \sum_{k=1}^{n} \pi_k <\phi_i|v|\phi_k>. \quad (13)$$

If this prescription is followed, if n = 1 for example, one simply recovers ordinary Rayleigh Schrodinger's Perturbation theory. The ramification of the use of MR-MBPT and ordinary MBPT have also been given by the group of Bartlett [8, 9].

Computational Considerations

In this, as in most other numerical studies, the algorithms chosen are designed first to achieve a desired level of precision and only then chosen to maximize efficiency. After all there is little value in achieving incorrect results, no matter how quickly. In these studies, we follow one of the conventional wisdoms of quantum chemistry and expand our orbitals in a basis set of gaussian orbitals. The primitive gaussian orbital is of the form:

$$X_{1\alpha\, i\, j\, k} (\vec{r} - \vec{R}_1) = (x^i y^j z^k / r^{i+j+k}) \exp(a_1 (\vec{r} - \vec{R}_1)^2) \quad (14)$$

This function has the advantage that all necessary integrals over these basis functions can be evaluated in closed form [10]. This allows one to know all integrals to arbitrary precision. Using this set, one may construct a "contracted" set of basis functions, j k, where

$$\xi_{1\, i\, j\, k} (\vec{r} - \vec{R}_1) = \sum_{\alpha} A_{\alpha 1}\, X_{1\, \alpha\, i\, j\, k} (\vec{r} - \vec{R}_1). \quad (15)$$

The A's in equation (15) are assumed given. One expands the Fock solutions in terms of these functions,

$$\phi_m^\beta = \sum_{i,\, j,\, k,\, l} C_{i\, j\, k\, l}^{m\beta}\, \xi_{1\, i\, j\, k} \quad (16)$$

The coefficients, $C_{i\, j\, k\, l}^{m\beta}$, are found using the Roothaan method, which is in reality a simple linear variation [11]. It is this we wish to discuss. The variation is performed by recalling for a particular iteration, the βth say,

$$\rho(\vec{x}\,\vec{x}') = \sum_{m=occ} \phi_m^{B+}(\vec{x}') \phi_m^B(\vec{x})$$

$$\equiv \sum_{ijkl} \sum_{i'j'k'l'} \xi_{1ijk}(\vec{x}) \xi_{1'i'\vec{x}_{j'}k'} \sum_m c_{ijkl}^{m\beta} c_{i'j'k'l'}^{m\beta*}.$$

The sum over m is restricted to occupied orbitals and mapping indices i j k l into an index i, and so forth, so that

$$\delta(\vec{x}\,\vec{x}') = \sum_{ii'} \xi_i(x) \xi_{i'}(x') S_{ii'}^B,$$

(17)

where

$$S_{ii'}^B = \sum_m c_i^{m\beta} c_{i'}^{m\beta*}$$

Thus only $S_{ii'}$ changes from iteration to iteration.

The Fock problem then reduces to a matrix problem (for each iteration) of the form

$$F\phi = E D \phi.$$

(18)

The matrices D and F are given as:

$$D_{ij} = \langle \xi_i | \xi_j \rangle,$$

(19)

and

$$F_{ij} = \langle \xi_i | F | \xi_j \rangle$$

$$\equiv \langle \xi_i | -\frac{\hbar^2}{2m}\nabla^2 - \sum_I \frac{e^2 Z_I}{|\vec{r} - \vec{R}_I|} | \xi_j \rangle$$

$$+ \sum_{kl} \Big[\langle \xi_i \xi_k | \frac{e^2}{|\vec{r}_1 - \vec{r}_2|} | \xi_j \xi_l \rangle$$

$$- \langle \xi_i \xi_k | \frac{e^2}{|\vec{r}_1 - \vec{r}_2|} | \xi_l \xi_j \rangle \Big] S_{kl}^B$$

(20)

Clearly the needed integrals merely need be evaluated once as they don't change from iteration to iteration. The only iteration dependent quantity is S_{kl}. Let us define;

$$f_{ij} = \langle \xi_i | -\frac{\hbar^2}{2m}\nabla^2 - \sum_I \frac{e^2}{|\vec{r} - \vec{R}_I|} | \xi_j \rangle$$

and

$$g_{ijkl} = \langle \xi_i \xi_k | \frac{e^2}{|\vec{r}_1 - \vec{r}_2|} | \xi_j \xi_l \rangle.$$

Then

$$F_{ij} = f_{ij} + \sum_{kl} \left[g_{ijkl} - g_{ilkj} \right] S^\beta_{kl}. \tag{21}$$

It is absolutely clear from equation (21) that each iteration is simply now a series of matrix operations followed by a matrix diagonalization.

The next step is to perform the correlation calculation. Consider the ordinary second order Rayleigh-Schrodinger case here. The second order correction to the energy is simply:

$$E^{(2)} = \sum_{i>j=occ.} \sum_{a>b=virt.} \frac{(V_{ijab} - V_{ijba})^2}{\epsilon_i + \epsilon_j - \epsilon_a - \epsilon_b}, \tag{22}$$

where

$$V_{ijab} = \sum_p C^{i\beta}_p \sum_r C^{j\beta}_r \sum_q C^{a\beta}_q \sum_s C^{b\beta}_s g_{pqrs}$$

$$= \langle \phi^\beta_i \phi^\beta_j | \frac{e^2}{|\vec{r}_1 - \vec{r}_2|} | \phi^\beta_a \phi^\beta_b \rangle \tag{23}$$

The coefficient C_p is the coefficient of the basis function in the i th Fock orbital for the βth configuration. Thus the dominent correlation problem becomes one of rotating integrals over basis functions to integrals over Fock orbitals. This again is simply a series of matrix-like steps. General considerations for programming equation (21), (22) and (23) may therefore be given.

The matrix of integrals over basis functions, gpqrs, is sparse (1% - 10% density). The sparseness has two causes, one being symmetry, the second being great separation of basis functions. Both considerations are taken into account before evaluating any integral, thus saving time. Once the integrals are generated one need be more particular in achieving efficiency. Consider as an example the part of the Fock matrix

$$F^{(2)}_{ij} = \sum_{kl} g_{ijkl} S^\beta_{kl} \tag{22}$$

or

$$F^{(2)}_I = \sum_K g_{IK} S^\beta_K \tag{22b}$$

In equation (22b) the indices i, j have been mapped into a single index I and k, l have been mapped into a single index K.

Computer type now enters into our consideration. In using vector computers like array processors, Cyber 205's, Cray's or in most cases. Therefore, the sparseness of the g matrix is of little help. It is of considerable help however, on conventional scalar-computers and one may work directly from equation (22) on such, loading only the non-zero integrals. On a vector machine however, one should use equation (22b) with the null integrals included, as each element of the Fock matrix is as seen a simple vector dot product. It is this operation which is maximally efficient on vector computers in general. Furthermore, the large length of the vector in (22b), typically of length 10^2 to 10^4, is ideal for such systems as the Cyber 205 as well as the other vector machines. Finally, this is in good form for processing on machines with parallel architecture such as a Denelcore or FPS164/MAX, as one can use the vector S_k as a constant and work on several of the F_I's at one time. Similar considerations apply to the matrix operations in equation (23).

Sample Results

This method was tested for the free Li^+ ion and for the Li^+-shell in Li. The ground state and the singlet and triplet 1s2s excited state were calculated in the UHF and the correlated limit. The results for the transition $1s^2$ to 1s2s (^3S) are 58.22 eV, 58.96 eV, and 59.01 eV in the UHF, the correlated and the experimental [12] limit respectively. The results for the transition $1s^2$ to 1s2s (^1S) 60.66 eV, 60.70 eV, and 60.75 eV in the UHF, the correlated and the experimental [12] limits respectively. These results indicate a significant ability for the present methods to make accurate calculation of the spectral energies. Projection operators are used to prevent excited states of the same symmetry as lower states from collapsing into these lower states, and to eliminate triplet contamination in the excited open shell singlet. These techniques are undergoing further development and will be the subject of a report when complete. The above test on the Li^+ free ion clearly indicates that correlation corrections are essential if one is to obtain accurate multiplet splittings. (The Unrestricted Hartree-Fock splitting has an error of 0.68 eV or 35%).

We now obtain the spectrum of the Li^+ ion in LiF using the LiF_6 (cluster embedded in a charge array for the ground state and the excited states corresponding to the $^{1,3}S$ and the $^{1,3}P$ states of the free Li ion. Using the BSW notation common to energy band theory we compute the $^{1,3}\Gamma_{15}$ states in the solid. In the UHF limit these states lie at 57.8, 61.2, 60.6, and 61.4 eV respectively for the $^3\Gamma_1$, $^1\Gamma_1$, $^3\Gamma_{15}$, and $^1\Gamma_{15}$ states respectively. Including correlation, these states lie at 58.0, 61.6, 61.5 and 62.3 eV respectively. The computed states lie several eV below the computed ionization limit for the Li^+ $1s^2$ state. There is no question as to the excitonic nature of these excitations. Since the ground state of the system is a $^1\Gamma_1$, the only one of these excited states which couples to the ground state by the dipole operator is the $^1\Gamma_{15}$ state. This lies at 62.3 eV

peak found near this energy. In general most theories and
experimental interpretations made recently are in agreement with
this point [13]. From here on there is no agreement to be found.
The relevant experiments all find a weak absorption feature at
about 61 eV, which appears as a shoulder on the 62.3 eV peak in
most data [13, 14]. The most recent interpretation given this
feature is due to Fields et al. who believe this to be the
transition to the $^1\Gamma_1$ state, that is, the 1s2s ^{1}S excited state
the Li. They base this comparison on the C. E. Moore data.
However, a word of caution is in order here. The 1s2s ^{1}S state of
the free Li$^+$ ion is not actually observed in nature. It is
possible that this state becomes optically accessible in a solid,
but an alternate possibility also suggested by the C. E. Moore
data is that one is seeing that the transition to the 1s2p ^{3}P
state which is found, albeit by indirect means, to lie at 61.27 eV
in the case of free Li$^+$. The energy for this transition is
computed to be 61.5 eV in the solid and is a possible candidate
for the shoulder. Energetics alone can not rule out the
transition to the 1s2s ^{1}S state as the above calculations
demonstrate. Likely this shoulder includes components of the $^1\Gamma_1$
and $^3\Gamma_{1,5}$ state.

There remains the 1s2s ^{3}S state to discuss. This state is
ignored by all the previous theories. As we have seen, this
transition lies at 58.0 eV in the solid according to our
calculation. The equivalent transition is predicted by these
calculations to lie at 58.96 eV in free space as opposed to 59.01
eV experimentally. There is every expectation that the solid
state value for this excitation may be accurate. The consequences
of this and estimate of transition strengths are given elsewhere
[15].

New Applications

The strength of these methods lies not only in the highly
accurate answers of which they are capable, but also in the
possible future developments of which they may be a part. After
all, the current calculations could be performed on a conventional
main frame or super mini in an acceptable time frame. There are
two proposed extensions for which the use of a super computer will
be essential, if one is to achieve answers in a finite time.
These shall be briefly described.

The first is a series of studies of the stability of various
charge states of point impurities in nonmetals. Due to the long
range coulomb potential, lattice ions may respond to a charged
impurity over a long distance. Both the ions will polarize and
the ionic positions may change as well. Currently, only classical
methods are being employed to study such effects in this detail
and then much essential electronic structure data is lost.
Current efforts by members of my group, that of John M. Vail at
the University of Manitoba, Canada, and of A. M. Stoneham at
Harwell, England are directed at solving this problem. The result
of this effort is a computer code named ICECAP which combines the
UHF + MBPT method described earlier with the lattice statics HADES
methods developed by the Harwell group. All this is achieved self
consistently and a current version of this code for a CRAY
computer exists. Development of this technique and application is
underway.

The second study made possible by the high efficiency of these methods and high speed computers is even more speculative. This arises from a collaboration of this group and that of G. Jacucci of the University of Toronto. This is an attempt to bypass the need for 2 body, etc. potentials in Molecular dynamic or Monte Carlo simulations of small clusters. Here, the total energy is to be evaluated in real time for a small cluster of atoms as part of a Monte Carlo simulation. Between 10^4 and 10^5 self-consistent calculations need be performed per simulation. This will limit application, but this study will provide a benchmark study, exact in some limit, by which to measure potential expansion methods.

REFERENCES

1) Mott, N. F. and Gurney, R. W.: Electronic Processes in Ionic Crystals (Dover, New York, 1940, 1948, 1964).
2) Fowler, W. B.: The Physics of Color Centers (Academic Press, New York, 1968).
3) Lannoo, M. and Bourgoin, J.: Point Defects in Semiconductors I, Theoretical Aspects (Springer-Verlag, Berlin, 1981).
4) Kunz, A. B. and Klein, D. L.: Phys. Rev. (1978) B 17, 4614.
5) Kunz, A. B.: in Theory of Chemisorption, (Springer-Verlag, Berlin, 1980).
6) Thouless, D. J.: The Quantum Mechanics of Many Body Systems (Academic Press, New York, 1961).
7) Davidson, E. R. and Silver, D. W.: Chem. Phys. Lett. (1977) 52, 403.
8) Bartlett, R. J., Shavitt, I. and Purvis, G. D.: J. Chem. Phys. (1979), 71, 281.
9) Lee, Y. S. and Bartlett, R. J.: (1984) to be published.
10) Boys, S. F.: Proc. Roy. Svc. (1950) A200, 542; ibid (1960) A258, 402.
11) Roothaan, C. C> J.: Rev. Mod. Phys. (1951) 23, 69; ibid (1960) 32, 179.
12) Moore, C. E.: Atomic Energy Levels I (National Bureau of Standards Circular 467, Washington D. C., 1949).
13) Field, J. R., Gibbons, P. C. and Schnatterly, S. E.: Phys. Rev. Lett. (1977) 38. 430.
14) Haensel, R., Kunz, C., and Sonntag, B., Phys. Rev. Lett. (1968), 20, 262.; Brown, F. C., Gahwiller, Ch., Kunz, A. B., and Lipari, N. O. Phys. Rev. Lett (1970) 25, 927.
15) Kunz, A. B., Boisvert, J. C. and Woodruff, T. O. Phys. Rev., in press.

Research supported in part by the U.S. Navy Office of Naval Research, Contract N00014-81-K-0620.

QUANTUM SIMULATIONS OF SMALL ELECTRON-HOLE COMPLEXES

Michael A. Lee
Department of Physics
Kent State University, Kent, Ohio 44242
R. Kalia and P. D. Vashishta
Division of Materials Science and Technology
Argonne National Laboratory, Argonne, Illinois 60439

ABSTRACT

The Green's Function Monte Carlo method is applied to the calculation of the binding energies of electron-hole complexes in semiconductors. The quantum simulation method allows the unambiguous determination of the ground state energy and the effects of band anisotropy on the binding energy.

INTRODUCTION

The interaction of electrons and holes in a semiconductor is apparently well described, but not exactly described, by the effective mass Hamiltonian. This simple Hamiltonian, which has a sound theoretic basis [1], says that electrons and holes interact via a coulomb potential which is screened by the dielectric constant. The fact that electrons in the conduction band and holes in the valence band interact with a sea of other electrons and a periodic lattice of atomic cores is contained in an effective mass m_e or m_h for the electrons and holes.

$$H = \Sigma - \frac{\hbar^2}{2m_e} \nabla^2_e - \frac{\hbar^2}{2m_h} \nabla^2_h + \Sigma \frac{e^2}{Kr_{eh}} \qquad (1)$$

With this Hamiltonian the study of small complexes becomes a few-body problem instead of a many-body problem. The simplest complex, the exciton, is equivalent then to a hydrogen atom and is an analytically soluble two-body problem. The existence of excitons is well known [2] from the luminesence spectra of semiconductors [3] and provide valuable information about the materials properties of semiconductors.

Small electron-hole complexes with two holes and one electron, the trion, with two holes and two electrons, the biexciton, and larger conglomerations have been the subject of theoretical consideration [4] with supporting experimental [5] evidence for some time. These few-body problems cannot be solved analytically, and the absence of an accurate ground state energy inhibits the unambiguous identification of their contribution to the spectra from these semiconductors.

It is the primary value of quantum simulation methods that they yield accurate, indeed exact, information about quantum many-body systems. This work describes the application of the Green's Function Monte Carlo (GFMC) method to the calculation of the ground state energy of these few-body electron-hole complexes. As will be explained in the next section, the method yields the exact (or as accurate a value as desired) ground state energy, but does not give an analytic expression for the wave function. All the information one obtains about the wave function comes in the form of a set of points in configuration space sampled from a probability distribution which is proportional to the wave function.

The GFMC Method

It has been appreciated for some time [6] that the mathematical structure of the Schroedinger equation is directly analogous to a diffusion equation in the presence of absorption processes. Kalos and co-workers [7] have taken this analogy and shown that it may be made into a computationally feasible method for obtaining exact information about the ground state energy and structure of quantum many-particle systems. To date, these methods and similar quantum simulation methods have been successfully applied to many-boson systems such as liquid and solid ^{4}He [8], one dimensional fermion models [9], few nucleon models [10] and lattice gauge theories [22]. Although no exact method exists [10] for simulating three dimensional many-fermion systems, considerable progress in this direction has been made for the electron gas [11], liquid ^{3}He [12], and atomic and molecular electronic structure calculations [13].

In this paper we will describe the Green's Function Monte Carlo method and apply it to the study of the four-particle (biexciton) system. Because this system has two electrons of opposite spin, and two holes of opposite spin, the antisymmetry requirement of fermion statistics does not explicitly enter the calculation, and normal boson methods apply.

The Diffusion Analogy

The Schrodinger equation for an N-particle system is

$$\{-\frac{\hbar^2}{2m} \sum_{i=1}^{N} \nabla_i^2 + V(\vec{r}_1, \ldots, \vec{r}_N)\} \psi = i\hbar \frac{\partial}{\partial t} \psi. \quad (2)$$

It will be convenient to use a 3-N dimensional vector, $R = (\vec{r}_1, \vec{r}_2, \ldots, \vec{r}_N)$, to specify the particle positions. To avoid confusion we will restrict the term particle to refer to the three dimensional entities in real space. In 3N dimensions the position vector R will specify a configuration. With this notation, and understanding that ∇^2 is the 3-N dimensional Laplacian, the Schroedinger equation becomes

$$-\frac{\hbar^2}{2m} \nabla^2 \psi(R,t) + V(R) \psi(R,t) = i\hbar \frac{\partial}{\partial t} \psi(R,t). \quad (3)$$

In the same notation, we wish to describe the diffusion process in 3-N dimensions. A density, $\rho(R,t)$, (to be thought of as the density of configurations having the coordinate R) gives rise to a diffusion current

$$\vec{J}(R,t) = -D\vec{\nabla}\rho(R,t) \tag{4}$$

where D is the diffusion constant. If current is not conserved, then a source term modifies the conservation equation,

$$\vec{\nabla}\cdot\vec{J}(R,t) + \frac{\partial}{\partial t}\rho(R,t) = S(R,t). \tag{5}$$

S(R,t) is the number of configurations per unit volume and time which are produced at point R. If S(R,t) is negative then its magnitude is the rate of absorption of particles. If there is a probability per unit time, A(R), for a particle at R to be absorbed, then the source term is of the form,

$$S(R,t) = -A(R)\rho(R,t). \tag{6}$$

Combining these three relations yields the diffusion equation in the presence of absorption.

$$-D\nabla^2\rho(R,t) + A(R)\rho(R,t) = -\frac{\partial}{\partial t}\rho(R,t). \tag{7}$$

It is the similarity of this equation to the Schroedinger equation which forms the basis of the GFMC method. In the appropriate units, ($\hbar=1$) the identifications of D with 1/2m and A(R) with V(R) make the two equations identical except for the factor $i = \sqrt{-1}$ associated with the time variable. This near equivalence can be seen clearly if at t=0 the starting point for the time evolution of both equations is the same, i.e. $\rho(R,0) = \psi(R,0)$, then the formal solution to the equations in terms of the stationary state solutions $\{\phi_n\}$ is

$$\rho(R,t) = \sum_n a_n \phi_n(R) e^{-\lambda_n t}$$
$$\psi(R,t) = \sum_n a_n \phi_n(R) e^{-iE_n t} \tag{8}$$

The eigenvalues are identical $\lambda_n = E_n$ and the eigenfunctions are identical. The difference is that ψ oscillates in time and ρ decays in time. Actually, the decay process occurs only if the eigenvalue spectrum is positive definite, but this can always be achieved by subtracting a constant E_T from the Hamiltonian. This amounts to guessing a trial value for the ground state energy. At large times, the dominate component in the density is the one with the smallest eigenvalue. If one has chosen E_T to be near E_0, then asymmptotically

$$\rho(R,t) \sim a_0 \phi_0(R) e^{-(E_0-E_T)t} \tag{9}$$

the density is nearly time independent and converges to the ground state wave function.

The crux of the GFMC method is that it doesn't attempt to simulate the Schroedinger equation directly, but actually implements a diffusion process that evolves in time until the density approaches the ground state wave function. Equivalently, one may say that the GFMC method simulates the quantum system in imaginary time. Following the second point of view, we define the variable $\tau = it$, then Schroedinger's equation

$$(-\nabla^2 + V(R) - E_T) \psi(R,\tau) = -\frac{\partial}{\partial \tau} \psi(R,\tau) \tag{10}$$

has the formal solution

$$\psi(R,\tau) = \sum_n a_n \phi_n(R) e^{-(E_n - E_T)\tau}. \tag{11}$$

The Green's Function

The operator which moves the system forward in imaginary time is the Green's function,

$$G = e^{-(H-E_T)\tau}. \tag{12}$$

In terms of position space variables, the propagation of ψ is achieved by the integral

$$\psi(R,\tau) = \int G(R,\tau,R',\tau') \psi(R',\tau') dR'. \tag{13}$$

It is this integral which is to be done using Monte Carlo techniques.

The properties of this Green's function are contained in elementary quantum mechanics and mathematical physics texts [14]. It is only necessary to make the substitution $\tau = it$ for the purpose at hand. G satisfies a diffusion equation

$$[-\nabla_R^2 + V(R)] G(R,\tau,R',\tau') = -\frac{\partial}{\partial \tau} G(R,\tau,R',\tau') \tag{14}$$

with the boundary condition

$$\lim_{\tau \to \tau'+} G(R,\tau,R',\tau') = \delta^{3N}(R-R'). \tag{15}$$

It has the formal solution

$$G = \Sigma \, \phi_n(R) e^{-E_n \tau} \phi_n(R') e^{+E_n \tau'} \tag{16}$$

and can itself be propagated forward in time

$$G(R,\tau,R'',\tau'') = \int dR' \, G(R,\tau,R',\tau') \, G(R',\tau',R'',\tau''). \tag{17}$$

In the language of a diffusion process, $G(R,\tau,R',\tau')$ is the density resulting from a unit source, or equivalently, the expected density at position R at time τ given that initially one configuration was at position R' at time τ'.

One does not need an analytic expression for the Green's function in order to propagate the wave function forward in time. It is useful, however, to begin with an analytic form which is valid for short times $\Delta\tau = \tau-\tau' \ll 1$. The configuration initially at R' remains localized there for short times. For sufficiently short times the local potential may be considered to be a constant, $u = V(R')$. The Green's function for a constant potential satisfies,

$$(-\nabla^2 + u) \, G_u(R,\tau,R',\tau') = -\frac{\partial}{\partial\tau} G_u(R,\tau,R',\tau')$$

and is given by a gaussian,

$$G_u(R,\Delta\tau,R',0) = \frac{\exp\{-(R-R')^2/2\Delta\tau - u\Delta\tau\}}{(2\pi\Delta\tau)^{3N/2}}. \tag{18}$$

In terms of $V(R')$, including now the constant E_T added to H, G is approximately given by

$$G(R,\tau'+\Delta\tau,R',\tau') \cong \frac{\exp\{-(R-R')^2/2\Delta\tau - (V(R')-E_T)\Delta\tau\}}{(2\pi\Delta\tau)^{3N/2}}. \tag{19}$$

and neglects terms of order $\Delta\tau$ compared to unity.

At this point one can implement the GFMC method within the short time approximation. We will come back to considerations of efficiency and accuracy after outlining the short-time algorithm.

The GFMC method will not give an analytic form for $\psi(R,\tau)$, but rather supplies a set of configuration points $\{R\}$ sampled from $\psi(R,\tau)$. That is to say the probability of a particular R occurring in a population of many configurations is proportional to the value of the wave function at that point. For the systems of interest here, the ground state wave function is positive definite, hence one can define the probability density

$$P(R,\tau) = \psi(R,\tau)/\int \psi(R,\tau) dR \tag{20}$$

and say that one obtains a set of configurations sampled from $P(R,\tau)$. Note that the probability density is in terms of the wave function and not its square.

The algorithm proceeds as follows. Initially one selects a large number ($M \sim 10^3$) of configurations $\{R_i(\tau=0)\}_{i=1}^{M}$ from some initial guess for the ground state wave function. This initial guess is usually an analytic trial wave function $\psi_T(R)$, and the initial sample can be generated using standard Metropolis techniques [15]. Given this sample for $\psi(R,0)$ one wishes to obtain configurations sampled from $\psi(R,\Delta\tau)$. The expected value of the wave function $\psi(R,\Delta\tau)$ at each point R is given by substituting the configurations $\{R_i(0)\}_{i=1}^{M}$ into equation 13. Then

$$\langle\psi(R,\Delta\tau)\rangle = \frac{1}{M} \sum_{i=1}^{M} G(R,\Delta\tau,R_i(0),0). \qquad (21)$$

If one samples a new set of L points $\{R_i(\Delta\tau)\}_{i=1}^{L}$ from the probability distribution

$$\langle\psi(R,\Delta\tau)\rangle/\int dR \langle\psi(R,\Delta\tau)\rangle, \qquad (22)$$

then this new set will be sampled from $P(R,\Delta\tau)$, i.e. the wave function at time $\tau = \Delta\tau$. Once complete, the process is repeated from the set $\{R(\Delta\tau)\}$ to obtain $\{R(2\Delta\tau)\}$ and integrated to arbitrarily large τ.

The simulation proceeds by allowing each of the initial configurations to diffuse for a time $\Delta\tau$. In the short time approximation G is known, and one simply sample a new set of configurations from the gaussians centered at the original configuration points. The short time Green's function is a normalized gaussian multiplied by a factor

$$W(R') = \exp(-(V(R')-E_T)\Delta\tau). \qquad (23)$$

This weighting factor is the probability that the configuration will survive a time $\Delta\tau$ in the presence of an absorption probability $V(R')-E_T$. If $W<1$, then with probability $1-W$, that configuration is eliminated from the simulation. Since $V(R')$ can be less than E_T, W may be greater than one. This is a branching ratio, or more precisely, the expected number of new configurations generated in a time $\Delta\tau$. Thus for $1<W<2$, a second configuration is sampled with probability $W-1$. For W greater than two, the generalization is obvious. After one has sampled 0, 1, or more configurations from each initial configuration, a new population of points has been obtained, and these points have been sampled from $\psi(R,\Delta\tau)$ (actually $P(R,\Delta\tau)$). This process has achieved one iteration of equation 13. The process is then repeated to obtain $\psi(R,2\Delta\tau)$ and may be iterated to arbitrarily large times.

Importance Sampling

Obtaining a population of points sampled from the ground state wave function is not enough to allow the calculation of expectation values. To state it simply, one cannot calculate the square of the wave function from the known set of configurations. Although the energy may be crudely obtained by adjusting the value of E_T so that the number of points in the population is stable, this estimate of E_0 has large statistical uncertainty. The random fluctuations in the size of the population due to the continual creation and annihilation of configurations can be largely eliminated through importance sampling. The philosophy at work here is the more information about the wave function which you put into the calculation, the more effective the GFMC method is in obtaining the results. The information is a reasonably good approximate analytic form for the ground state wave function $\psi_T(R)$. This is typically a Jastrow type wave function which incorporates some short range pair correlations, a feature particularly important in coulomb systems, or systems interacting with unbounded potentials.

Using the trial wave function $\psi_T(R)$, one forms the new density,

$$f(R,\tau) = \psi(R,\tau) \psi_T(R) \tag{24}$$

and the new propagator

$$K(R,\tau,R',\tau') = \psi_T(R) G(R,\tau,R',\tau')/\psi_T(R')$$

which then satisfies the modified propagator equation,

$$f(R,\tau) = \int K(R,\tau,R',\tau') f(R',\tau')dR'. \tag{25}$$

This equation says that if a set of configurations is propagated forward in time according to the modified stochastic dynamics of the kernal K, then this new population eventually evolves to

$$\lim_{\tau \to \infty} f(R,\tau) \propto \phi_0(R) \psi_T(R). \tag{26}$$

Before discussing the modified propagation procedure and its effect on population stability, it is pointed out the secondary advantage of using $f(R,\tau)$ in the calculation of the energy. If we calculate the expected value of $(H\psi_T(R))/\psi_T(R) \equiv E(R)$ over M configurations the expression is

$$\langle (H\psi_T)/\psi_T \rangle = (1/M) \sum_{i=1}^{M} H\psi_T(R_i)/\psi_T(R_i). \tag{27}$$

This is an estimator for the expectation value over the probability density

$$P(R,\tau) = f(R,\tau)/\int dR f(R,\tau). \tag{28}$$

The average equals the ground state energy when $f(R) \propto \phi_0(R) \psi_T(R)$. The expectation value

$$\langle \frac{H\psi_T}{\psi_T} \rangle = \frac{\int dR \phi_0(R)\psi_T(R) [H\psi_T(R)/\psi_T(R)]}{\int dR \phi_0(R)\psi_T(R)} \tag{29}$$

$$= E_0 = \frac{\int dR \phi_0(R) H\psi_T(R)}{\int dR \phi_0(R)\psi_T(R)} \tag{30}$$

by the hermitian property of H. It can be shown [12] that the expectation value of E(R) is an upperbound to E_0 at any τ. The energy calculated in this fashion has a much lower statistical variance if an accurate ψ_T is known. Indeed, the statistical uncertainty would be zero if $\psi_T = \phi_0$.

The difference in the propagators K and G can be made clear if one expands the ratio $\psi(R_2)/\psi(R_1)$ using

$$\ln \psi(R_2) \cong \ln \psi(R_1) + (R_2-R_1)\cdot \frac{d}{dR_1}\ln \psi(R_1) + \frac{1}{2}(R_2-R_1)^2 \frac{d^2}{dR_1^2}\ln \psi(R_1). \quad (31)$$

Then, K is approximately given by

$$K = G(R,\tau,R',\tau') \exp[\ln\psi_T(R) - \ln\psi_T(R')] \quad (32)$$

$$\cong (1/2\pi\Delta\tau^*)^{3N/2}\exp\{-(R-(R'-F\Delta\tau^*))^2/2\Delta\tau^* - (E(R')-E_T)\Delta\tau^*\}$$

where $F = \vec{\nabla}'\ln\psi(R')$ and $\Delta\tau^* = \Delta\tau/(1-\nabla^2\ln\psi(R'))$.

This expression is correct to order $\Delta\tau$ as before, and we have neglected some cross terms of the order $(R-R')^2$ in the Taylor series expansion. This is consistent since $(R-R')^2 \sim \Delta\tau$. In this form, we see that if the trial function ψ_T is exactly equal to ϕ_0, then $E(R) = E_0$ and provided $E_T = E_0$, the propagator is a normalized gaussian. Hence the size of the population never changes and all configurations survive each step with probability W=1. When ψ_T is close to ϕ_0, the population fluctuates, but the fluctuations are much smaller. The dynamics of the diffusion have been changed by translating the center of the gaussian by an amount $F\Delta\tau^*$ towards regions of larger ψ_T, i.e. higher probability. The width of the gaussian has also changed. The population stability has been gained but at the expense of calculating E(R) at each step rather than V(R). The overall result is still a much more efficient, lower variance calculation.

The Biexciton

The biexciton in its ground state consists of two electrons of opposite spin and two holes of opposite spin. The existence of the entity is without question, but its binding energy has been debated [16] and even accurate variational calculations [16] have given only 50% of the experimental binding energy in silicon and germanium. From the variational calculations of Brinkman, Rice, and Bell [16] we know that wave functions which incorporate some particle correlations yield improved binding energies. One might expect then that a GFMC calculation, which incorporates all correlation effects, would obtain significantly greater binding energies and resolve the discrepancies with experimental results.

To eliminate uncertainties in materials parameters it is best to establish a system of units where energy is measured in units of the exciton energy, E_x, given by

$$E_x = \mu e^4/2\hbar^2\kappa^2 \quad (33)$$

where μ is the reduced mass and K the dielectric constant. In these units, the four-particle Hamiltonian becomes

$$H = -(1/(1+\sigma))(\nabla_1^2 + \nabla_2^2) - (\sigma/(1+\sigma))(\nabla_a^2 - \nabla_b^2)$$
$$+ 2/r_{12} + 2/r_{ab} - 2/r_{1a} - 2/r_{1b} - 2/r_{2a} - 2/r_{2b} \qquad (34)$$

where, $\sigma = m_e/m_h$, and the electron coordinates are labeled 1,2 and the hole coordinates a,b.

The trial wave function, $\psi_T(R)$ was the product of three functions ψ_{ee}, ψ_{hh} and ψ_{eh}, chosen to incorporate as much information about pair correlations as possible.

$$\psi_{ee}(r) = \exp[c_1 r/(1+c_2 r)]$$
$$\psi_{hh}(r) = \exp[c_3 r/(1+c_4 r)]$$
$$\psi_{eh} = \exp[-(\alpha r_{1a} + \beta r_{1b} + \beta r_{2a} + \alpha r_{2b})]$$
$$+ \exp[-(\beta r_{1a} + \alpha r_{1b} + \alpha r_{2a} + \beta r_{2b})] \qquad (35)$$

The GFMC calculations were performed for several values of the mass ratio, σ = 0.01, 0.1, 0.3, 0.6 and 1.0. The variational parameters c_1, c_2, c_3, c_4 and α and β were varied until a reasonable initial energy was obtained at each σ. Populations of ≈500 configurations were run for typically 40 units of time with time steps of $\Delta\tau$ = .005 or less. Tests were done on time steps of $\Delta\tau$ = .01 and .001 to establish that the error due to the short time approximation was less than the statistical uncertainty (<0.1%) of the total energy.

Figure 1 shows the results from the GFMC calculations and the results of the variational calculations of ref. 16. In these units, the energy of two isolated excitons is -2.0. One sees that as the electron and hole masses become comparable the system becomes weakly bound. When σ = 1, the binding energy is about 3% of the total energy. It is in this equal mass limit that the variational calculations suffer most from inadequate treatment of correlations in the wave function, and yield a binding energy of only half the correct value. The 1-2% error in the total variational energy becomes less important as the hole mass increases.

Comparison of the GFMC results to experimental measurements can be made for several values of σ. For small σ an exact form for the energy is known [17],

$$E(\sigma) = (-2.346 + 0.764\sqrt{\sigma})E_x \qquad (36)$$

and agrees with the GFMC results at σ = .01 to better than three significant figures. In this mass range CuBr and CuCl [2,18] have values of σ = 0.01 and 0.02 and experimental binding energies of ≈29 mev and 34-44 mev respectively.

The exciton energies of 110 and 190 mev in these systems give GFMC energies of 29 and 45 mev respectively. This is very good agreement, but it must be treated cautiously since a small uncertainty in the mass makes a big change in the energy in this region of small σ. A more reasonable comparison may be to take the measured binding energy and predict the mass ratio, since it is the less accurate quantity.

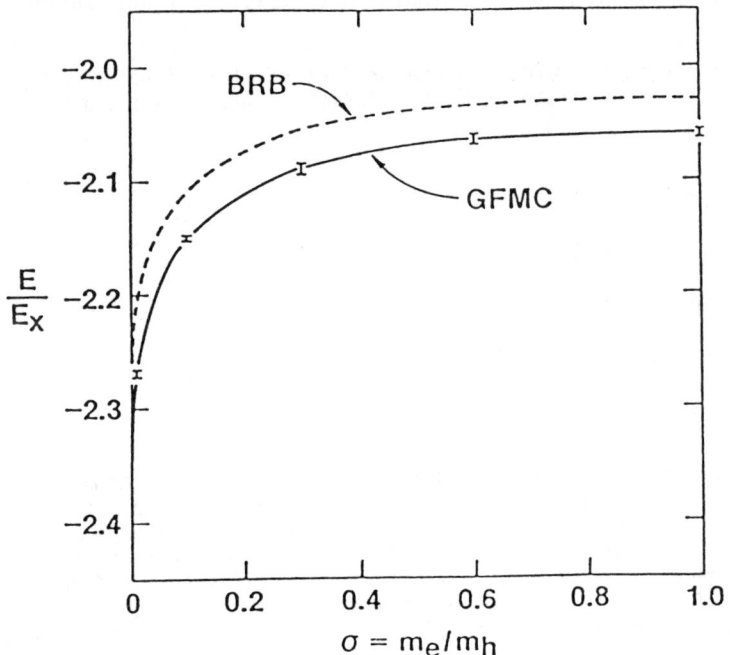

Figure 1. Ground-state energy of biexcitons as a function of the electron-hole ratio, σ. The dotted curve represents the variational results of Brinkman, Rice and Bell (BRB) (ref.16) while the solid curve shows the Green's Function Monte Carlo (GFMC) results. Here the energies are expressed in excitonic rydbergs, E_x.

Germanium and silicon are well studied systems and have electron-hole masses which are more nearly equal. There is, however, the complication of band warping since the top of the two valence bands in these semiconductors are degenerate and coupled. Recently biexcitons have been observed stressed Ge<1,1,16> [19] and Si<1,0,0> [3] where the electrons occupy a single conductio band and the holes occupy a single hole band. These bands are anisotropic, but we will come back to that point later. In the germanium experiments the binding energy is estimated to be .15 ± .01 mev and using σ ≈ .7, the GFMC energy corresponds to a .16 mev binding energy. The variational calculation obtains only 60% of this binding. In the experiments on stressed Si<100>, Gourley and Wolfe [3] report the binding energy to be 0.10 E_x which is to be compared with a value of 0.08 E_x in unstressed Si [20] and a GFMC value of .06 E_x when a mass ratio of $\sigma \approx 1$ is taken.

It is only in silicon that there appears to be a discrepancy between calculated and experimental results. The variational result accounts for about 1/3 of the binding and the GFMC accounts for about 2/3 of the observed binding energy. This is one of the advantages of an exact numerical result. A discrepancy with experiment tells you something, because it cannot be attributed to approximations invoked in arriving at the solution of the problem. In the present case we have essentially an exact result for the ground state energy of the effective mass Hamiltonian. Apparently, this Hamiltonian does not exactly correspond to the experimental system. One difference is that the calculation has not included band anisotropy. Because the curvature of both the electron and hole bands is different along different crystal axes, the effective masses along these axes must be distinguished as to the longitudinal m_L and transverse m_T components. Returning for a moment to the analogy between the Schroedinger equation and a diffusion equation, this is equivalent mathematically to saying that the system has an anisotropic diffusion constant. This feature makes even the two-body exciton Hamiltonian insolvable analytically but causes only a minor change in the GFMC calculation. We have accordingly taken literature values [21] of the anisotropic masses in silicon, and repeated the silicon calculations to determine the effect of this anisotropy. Preliminary results indicate that at most the binding energy is lowered another 10% removing perhaps a third of the remaining discrepancy. Compared to the total energy this is a small difference $\approx 1.5\%$. One may even take this result as support for the surprising accuracy of the effective mass Hamiltonian. Alternatively one may use it to motivate an attempt to find the physical origins of the additional binding energy.

REFERENCES

1) Sham, L.J. and Rice, T.M.: Phys. Rev., 1966, 144, 708.
2) Rashba, E.I. and Sturge, M.D.: Excitons (North Holland, New York, 1982).
3) Wolfe, J.P.: Physics Today, 1982, 35, 46; Gourley, P.L. and Wolfe, J.P.: Phys. Rev. B, 1979, 20, 3319; Reynolds, D.C. and Collins, T.C., Excitons, (Academic Press, New York, 1981).
4) Wang, J.S.Y. and Kittel, C.: Phys. Lett. A, 1972, 42, 184; Rice, T.M. in Solid State Physics, ed. Ehrenreich, H., Seitz, F. and Turnbull, D. (Academic Press, New York, 1977).
5) Kulakovskii, V.D. and Timofeev, V.B.: Pis'ma Zh. Eksp. Teor. Fiz., 1977, 25, 487 (trans. JETP Lett. 1977, 25, 458); Gourley, P.L. and Wolfe, J.P.: Phys. Rev. Lett., 1978, 40, 526; Hensel, J.C., Phillips, T.G. and Thomas, G.A. in Solid State Physics, ed. Ehrenreich, H., Seitz, F. and Turnbull, D. (Academic Press, New York, 1977).
6) Metropolis, N. in Symposium on Monte Carlo Methods, ed. Meyer, H. (John Wiley, New York, 1956), p. 24.
7) Kalos, M.H., Levesque, D. and Verlet, L.: Phys. Rev. A, 1974, 9, 2178; Ceperly, D.M. and Kalos, M.H., in Monte Carlo Methods in Statistical Physics, ed. Binder, K. (Springer, Berlin, 1979), pp. 145-197.
8) Kalos, M.H., Lee, M.A., Whitlock, P.A. and Chester, G.V.: Phys. Rev. B, 1981, 24, 115.
9) Lee, M.A., Motakabbir, A.K. and Schmidt, K.E.: Phys. Rev. Lett., 1984, 53, 1191.

10) Arnow, D.M., Kalos, M.H., Lee, M.A. and Schmidt, K.E.: J. Chem. Phys., 1982, 27, 5562.
11) Ceperly, D.M. and Alder, B.J.: Phys. Rev. Lett., 1980, 45, 566.
12) Lee, M.A., Schmidt, K.E., Kalos, M.H. and Chester, G.V.: Phys. Rev. Lett., 1981, 46, 728.
13) Moskowitz, J.W., Schmidt, K.E., Lee, M.A. and Kalos, M.H.: J. Chem. Phys., 1982, 77, 349; Ceperley, D.M., Alder, B.J. and Lester, W.A.: J. Chem. Phys., 1982, 77, 5593.
14) Morse, M.M. and Feshbach, H., *Methods of Theoretical Physics*, (McGraw-Hill, 1953).
15) Metropolis, N., Rosenbluth, A.W., Rosenbluth, M.N., Teller, A.M. and Teller, E.: J. Chem. Phys., 1953, 21, 1078.
16) Brinkman, W.F., Rice, T.M. and Bell, B.: Phys. Rev. B, 1973, 8, 1570.
17) Rehner, R.K.: Solid State Commun., 1969, 7, 457.
18) Grun, J.B., Nikitine, S., Bivas, A. and Levy, R.: J. Lumin., 1970, 1, 241; Souma, H., Gota, T., Ohta, T. and Ueta, M.: J. Phys. Soc. Jpn., 1970, 29, 697; Mysyrowicz, A., Grun, J.B., Levy, R., Bivas, A. and Nikitine, S.: Phys. Lett., 1968, 26A, 615; Gato, T., Souma, H. and Ueta, M.J.: Lumin., 1970, 1, 231.
19) Kukushkin, I.V., Kulakovskii, V.D. and Timofeev, V.B.: JETP Lett., 1980, 32, 281.
20) Thewalt, M.L.W. and Rostoworowski, J.A.: Solid State Commun., 1978, 25, 991.
21) Vashishta, P.D., Kalia, R.K. and Singwi, K.: in *Electron-Hole Droplets in Semiconductors*, ed. C.D. Jeffries and L.V. Keldysh, (North Holland, 1983).
22) Heys, D.W. and Stump, D.R.: Phys. Rev. D, 1983, 28, 2067.

THEORY OF SUPERCONDUCTING ARRAYS IN A MAGNETIC FIELD

D. Stroud and W. Y. Shih
Department of Physics
Ohio State University
Columbus, Ohio 43210

ABSTRACT

We review the theory of superconducting arrays in a magnetic field. The arrays are assumed to consist of grains of superconductor embedded in a non-superconducting host, and coupled together by the proximity effect or Josephson tunneling. In the presence of a field, ordered two-dimensional arrays are found to undergo a coherent/incoherent transition at a temperature $T_c(B)$ below which the phases of the superconducting order parameters on different grains become coherent over macroscopic distances, but the temperature at which this transition occurs is violently field-dependent. The transition is analyzed both by mean-field theory and by Monte Carlo simulation. Similar effects are found in three-dimensional ordered samples, but disordered arrays in either two or three dimensions are found to behave very much like a spin glass at strong enough fields, having only a glass transition below which the phases are frozen but not ordered, as in more usual glasses. Experiments on two-dimensional triangular arrays are found to confirm theoretical predictions in some detail, but comparison with predictions for disordered arrays has not yet been made.

INTRODUCTION

One of the most characteristic features of bulk superconductors is their response to an external magnetic field. Type I superconductors expel flux at temperatures below the superconducting transition temperature T_c, except for a thin surface layer of temperature-dependent thickness $\lambda(T)$ (the penetration depth); they are driven normal at a critical magnetic

 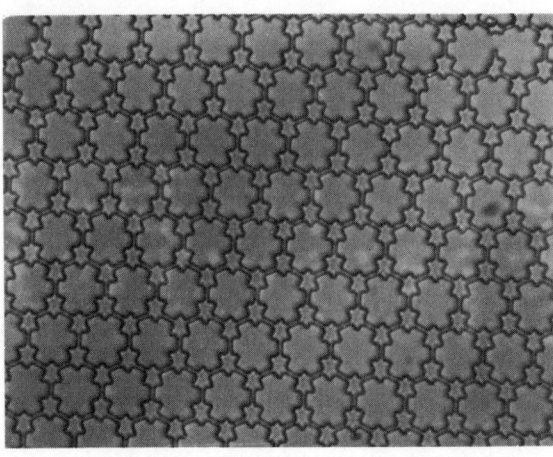

Figure 1. Examples of superconducting arrays: photographs of "disk" and "asterisk" triangular arrays used in the experiments of Brown, Rudman, and Garland. Both are Pb (disks or asterisks) in a matrix of Sn, and placed on a triangular lattice of lattice constant 15 μ (disks) and 10 μ (asterisks). In both cases, thickness of the film is about 1500 A. Disk arrays were made photomicrolithographically at the University of Cincinnati; asterisk arrays were made at the Cornell Submicron Facility.

field $H_c(T)$. Type II superconductors behave like Type I up to a lower critical field $H_{c1}(T)$, above which the applied magnetic field penetrates in the form of vortices but the material retains zero resistivity. At the upper critical field $H_{c2}(T)$, flux penetrates completely and the material returns to its normal state. The behavior especially at type I superconductors was historically very important in helping theorists formulate successively more microscopic accounts of the superconducting state culminating with the BCS synthesis in 1957.

In recent years, there has been increasing interest in inhomogeneous superconductors [1, 2]. Such materials are typically composed of two constituents, one of which is a superconductor, while the other is a normal metal, semiconductor, insulator, or superconductor with a lower transition temperature. The interest results from the many unusual properties of these materials, from the fact that they can be prepared in variety of different ways ranging from photomicrolithographic deposition with a precisely controlled microstructure to various kinds of random stirring, and possibly from the practical technological potential of these materials in, for example, applications involving high current and magnetic field and density, Josephson junction computers.

In this paper, we shall describe a theory to treat the magnetic properties of one class of superconducting composites, which may be called superconducting arrays (figure 1). We use this term rather loosely to refer to a collection of superconducting particles embedded in a non-superconducting host and couple together by Josephson tunneling or by the proximity effect. Such an array may have an ordered or disordered geometrical arrangement [Examples of the first are illustrated in figure 1.] An array may be one-, two- or three-dimensional. One- and two-dimensional arrays can be readily prepared in ordered form by microfabrication techniques, but three-dimensional "arrays" will generally be disordered (though ordered ones are imaginable). The superconducting particles themselves can also be d-dimensional with d ranging from zero to three. By "d-dimensional" we mean that the particle is large compared to other characteristic lengths of the problem in d directions out of three. The important characteristic lengths are λ, the penetration depth; and $\xi(T)$, the superconducting coherence length of the superconductor (that is, roughly speaking, the size of the Cooper pair). With this range of possibilities, the possible behavior of these arrays in a magnetic field obviously has much variety.

We shall be concerned here principally with two- and three-dimensional arrays, both ordered and disordered. The dicussion will concentrate on the magnetic properties arising from weak coupling between the grains, and not from the grain diamagnetic response itself. Effects due to coupling will obviously dominate when the grains are zero-dimensional, that is, are quite small (less than 1000 A or so for real materials of interest), but even for larger grains such as are seen in many experiments, it is often possible in principle, and sometimes in practice, to separate effects due to coupling from single-grain phenomena. Isolated-grain effects have, in any case, been well treated elsewhere [3].

Although weak-coupled arrays are emphasized here, many phenomena observed in such arrays can also be seen in an apparently quite different system consisting of superconducting wire networks [4]. Like the arrays, such networks can be prepared in ordered or disordered form by photolithographic techniques and other methods. If the wires are thin compared to a penetration depth, it can be shown that they will behave in many respects like the arrays in a magnetic field. For reasons of space, these networks will not be further analyzed in this paper.

As a further apologia, it should be noted that this paper is intended only as a survey of some interesting phenomena that can be predicted for superconducting arrays in a magnetic field. It is not a comprehensive review of work on the field, but instead will emphasize theoretical work at Ohio State, only because the author is most familiar with this work. Representative references to other work are however, provided to the extent possible.

The remainder of this paper is arranged in the following way. Section II describes a model which is useful in elucidating the behavior of the arrays in a magnetic field, and outlines the qualitative behavior to be expected from this model in various geometries, as well as the limitations of the model. Section III presents applications of the model to ordered arrays, primarily two-dimensional arrays, since these are the materials which can be prepared experimentally with greatest ease. Applications to disordered arrays are outline in Section IV. Section V summarizes various experiments which provide tests for these models, and offers some concluding remarks.

II. Model

A bulk superconductor, as is well known, is characterized by a complex order parameter $\psi = \Delta\exp(i\phi)$. The magnitude $|\psi| = \Delta$ is, in general, temperature-dependent, and is proportional to the BCS energy gap of the superconductor. If no current flows, and no vector potential is present, and if the geometry is such that Δ is spatially uniform, the Helmholtz free energy of the superconductor depends only on the magnitude of the order parameter, and not on its phase, ϕ. The phase of the order parameter is thus a thermodynamic variable which may be varied to give a continuous sequence of degenerate thermodynamic states.

As was first pointed out by Josephson, the phase becomes a dramatically more relevant variable when one considers two superconductors in close proximity so that a current of Cooper pairs can flow between them. If the two superconductors have spatially uniform order parameters $\psi_1 = \Delta_1\exp(i\phi_1)$ and $\psi_2 = \Delta_2\exp(i\phi_2)$, then there exists a term in the free energy which depends on the phase difference between the superconductors, namely

$$E_J = -J_{12}\cos(\phi_1 - \phi_2) \ . \tag{1}$$

The coupling energy is related to the maximum, or critical, supercurrent I_{12} that can flow between the two superconductors by the equation

$$J_{12} = \frac{h}{2e} I_{12} \tag{2}$$

where h and e are fundamental constants. The phases in equation 1 behave like thermodynamic variables, and their average is to be computed within an appropriate thermodynamic ensemble, i.e. the canonical ensemble at finite temperature.

Equation 1 can be generalized in an obvious way to array of superconductors weakly coupled together by Josephson tunneling (or the proximity effect, if the medium between the superconducting grains is a normal metal). The appropriate generalization involves a sum over all distinct pairs $\langle ij \rangle$:

$$H = -\sum_{\langle ij \rangle} J_{ij}\cos(\phi_i - \phi_j) \ , \tag{3}$$

where ϕ_i is the phase of the order parameter on the ith grain, and by using the symbol H for the interaction energy we emphasize the fact that the right hand side represents an interaction Hamiltonian which is to be averaged with respect to the classical canonical ensemble to give the measurable equilibrium properties of the array at finite temperature. All such properties may be derived from knowledge of the partition function,

$$Z = \int\int\int...\exp[-H(\phi_1,...)/k_BT]d\phi_1 d\phi_2... \tag{4}$$

and corresponding free energy $F = -k_BT \ln Z$.

Much insight can be gained from the observation that the Hamiltonian H is formally identical to that of classical two-component ("x-y") spins of fixed length Δ_i and direction in the x-y plane specified by the angle ϕ_i. Since each spin represents a macroscopic quantum condensate, the Hamiltonian refers to a much smaller number of variables than the typical 10^{23} of microscopic problems. Nonetheless, the universality classes of possible phase transitions are the same in both cases. In particular, a two-dimensional array is predicted to undergo the unusual quasi-continuous phase transition discussed by Kosterlitz and Thouless [5] and by Berezinskii, [6] and various experiments have indeed given some evidence that such a transition occurs in arrays of weakly coupled superconductors [2].

In the presence of a magnetic field, the phase difference appearing in equation 3 is replaced by the generalization of this quantity to reflect a non-zero vector potential. The new Hamiltonian is thus

$$H = -\sum_{\langle ij \rangle} J_{ij} \cos(\phi_i - \phi_j - A_{ij}) \qquad (5)$$

where A_{ij} is given by the line integral of the vector potential $\vec{A}$ from the center of grain i to that of grain j:

$$A_{ij} = \frac{2\pi}{\Phi_o} \int_i^j \vec{A} \cdot d\vec{\ell} \qquad (6)$$

In equation 6 Φ_o = hc/2e is a flux quantum. The presence of this new phase factor totally alters the physics of the system of interacting grains. This is easily seen by considering the form of the phase factors A_{ij} for a particular choice of gauge. If we consider a magnetic field in the z direction, say $\vec{B} = B\hat{z}$, then a convenient gauge is

$$\vec{A} = Bx\hat{y} \ . \qquad (7)$$

In this gauge, bonds parallel to the x and z axes are unaffected by the magnetic field, whereas for a bond in the y direction the phase factor is

$$A_{ij} = 2\pi Bxa/\Phi_o \qquad (8)$$

where a is the length of the bond. As B is varied, the phase factor increases continuously; for large enough B, it may be many times larger than 2π. Thus, if we continue the pseudospin analog, the interaction between spins i and j along the y axis varies continuously from "ferromagnetic" to "antiferromagnetic" and back again, as B is increased. The types of phase transitions in arrays described by this Hamiltonian, and the corresponding resistive behavior, is obviously very complex, and forms the chief subject of the remainder of this paper.

Before turning to a detailed discussion of this behavior, it is worth pointing out which (potentially important) physical phenomena are not included in the Hamiltonian, equation 5. One omission is the so-called charging energy, which is a term in the Hamiltonian arising from finite intergrain capacitance. In its most general form, this term may be written [7].

$$H_C = \frac{1}{2} \sum_{\langle ij \rangle} U_{ij} n_i n_j$$

when n_i is the excess charge on grain i, in units of an electron charge, and U_{ij} is proportional to the inverse capacitance matrix for the array of superconducting grains. If it is assumed that charge transfer between grains occurs only by tunneling of Cooper pairs, then n_i is proportional

to the number operator for Cooper pairs on site i. It is therefore
quantum-mechanically conjugate to the phase variable on site i, and the
pair (n_i, ϕ_i) obeys a Heisenberg uncertainty relation. Large values of
U_{ij} (typically associated with small grains) favor small values of n_i, that
is, charge-neutral grains, and hence, by the uncertainty relation, no fixed
relation between the phases on different grains. In other words, large
charging energies tend to suppress phase coherence. This effect may be
quite important in arrays made of tunnel junctions [8], where capacitances
may be low, but are less important in proximity effect arrays [9-11]. The
precise mathematical treatment of the Hamiltonian when both phase-ordering
and charging energies are important is quite difficult and remains a matter
not fully resolved.

Equation 5 also makes the oversimplified assumption of "point grains."
Of course, the grains (and the junctions) have finite extent. The vector
potential therefore varies from point to point within the junction, and the
expression given in equation 5 for the coupling must be modified. Since
the exact modification depends in detail in the geometry of the junction,
as well as on assumptions about how far the external field penetrates the
grains, and the degree to which it differs from the local field, it is very
difficult to improve on this approximation in any reliable way.

III. Ordered Arrays

Many of the most striking features of the Hamiltonian of equation 5 can
be seen when it is applied to two-dimensional ordered arrays of grains, a
geometry which is also very convenient to prepare experimentally. The
utility of this model was first pointed out by Teitel and Jayaprakash in a
pair of seminal papers of 1983 [12]. Since their work, there have been
several additional papers on similar geometries, some of which are
discussed below.

In a two-dimensional lattice of grains, it is readily shown that the
Hamiltonian, equation 5, is periodic with a period of one flux quantum per
unit cell of the grain lattice (and hence, all equilibrium properties of
the lattice described by this Hamiltonian are also periodic). This is most
easily seen for the square lattice with bonds parallel to the x and y axis,
where a simple substitution of the vector potential, equation 7, into
equation 5 demonstrates periodicity, but it is true for other lattice
structures also, such as the triangular lattice which is often studied
experimentally. In the honeycomb lattice, there are two grains per unit
cell, and the unit cell has area of one elementary hexagon. In this case,
the model is periodic with a period of one flux quantum per hexagon.

The periodicity is most easily understood in terms of a concept known
as "frustration," first introduced by Toulouse [13] to describe magnetic
systems with competing ferromagnetic and antiferromagnetic interactions.
For the present model, the frustration around any closed plaquette of
connected bonds is defined by

$$f = \frac{1}{2\pi} \sum A_{ij} = \Phi/\Phi_0 \tag{9}$$

where $\sum$ denotes the sum of the phase factors A_{ij} around the plaquette, and
Φ is the magnetic flux through the plaquette by Stokes' theorem. Whenever

all the factors f are integers, it is always possible to make a change of phase variables. For non-integer f, no such transformation is possible, and the Hamiltonian always has the property that some of the bonds tend to align the phases parallel (i.e. are "ferromagnetic") while other bonds favor non-parallel phases. For such a Hamiltonian, there exists no choice of phases which simultaneously minimizes all the bond energies, and the system is therefore justly described as "frustrated". For a periodic lattice with nearest neighbor interactions, the frustration is homogeneously distributed throughout the lattice ("uniform frustration").

The Hamiltonian of equation 5 is very difficult to analyze in the presence of frustration. Because of this difficulty, it is useful to have some rough method of estimating its behavior. Such a method is provided by molecular field theory, first worked out for this problem by Shih and Stroud [14]. While molecular field theory is certainly not accurate for two-dimensional systems, and may even given misleading results for them, nonetheless it remains a valuable standard against which more exact calculations may be measured. In the present case, several of the predictions of molecular field theory appear to be confirmed by experiment, as discussed below.

The molecular field theory begins by defining order parameters

$$n_i = \langle \exp(i\phi_i) \rangle \tag{10}$$

where the brackets denote a canonical average given by

$$\langle \exp(i\phi_i) \rangle = \frac{1}{Z} \int ... \int d\phi_1 ... d\phi_N \exp(i\phi_i) \exp(-H/k_B T) \tag{11}$$

where Z is given by equation 4. This exact expression is replaced, in molecular field theory, by the approximation

$$n_i \sim Z_i^{-1} \int_0^{2\pi} d\phi_i \exp(i\phi_i) \exp(-\beta H_{eff}^{(i)}) \tag{12}$$

$$Z_i = \int_0^{2\pi} d\phi_i \exp(-\beta H_{eff}^{(i)}) \tag{13}$$

$$H_{eff}^{(i)} = -\sum_{j \neq i} J_{ij} [\cos\phi_i \langle \cos(\phi_j + A_{ij}) \rangle + \sin\phi_i \langle \sin(\phi_j + A_{ij}) \rangle] \tag{14}$$

where the expectation values in equation 14 are to be evaluated self-consistently from equations of the form 12. Equations 12-14 form a typical set of self-consistent equations analogous to those found in other types of magnetic problems, such as the Weiss model of ferromagnetism or antiferromagnetism. In the present case, they constitute (for N grains) a set of N coupled complex non-linear equations, or 2N real equations, for the N unknown complex phase order parameters n_i, and in general must be solved numerically. In principle, they are applicable to three-dimensional as well as two-dimensional arrays, and to disordered as well as ordered ones.

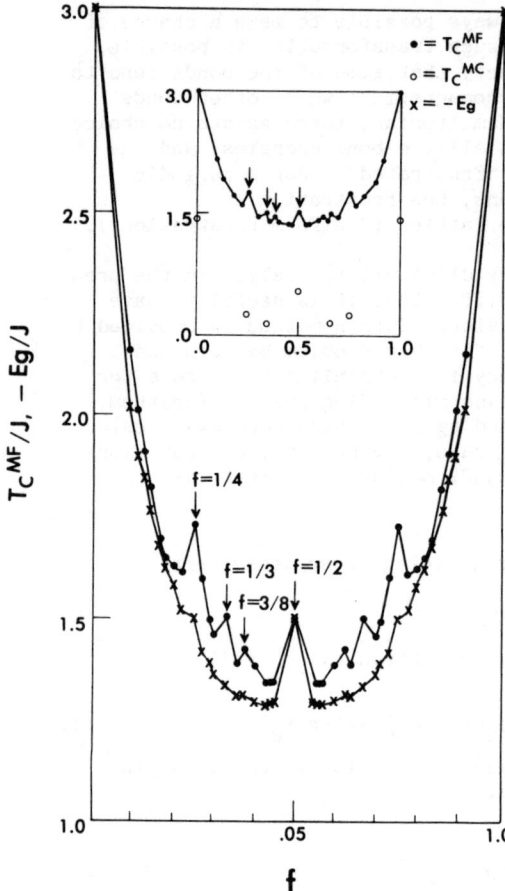

Figure 2. Mean field transition temperature (dots connected by solid line), ground state energy (crosses connected by solid line), and Monte Carlo transition temperature (open circles) for the triangular lattice, as a function of field. f is the field in units of flux quanta per unit triangle. The solid lines are merely to guide the eye. From W.Y. Shih, Ph.D. Dissertation, Ohio State (unpubl.).

The quantity of greatest interest in mean-field theory is, of course, the mean-field phase transition temperature $T_c(B)$ (or equivalently, $T_c(f)$, where f now represents the fraction of magnetic flux per unit cell of the grain lattice). This is the temperature at which mean-field theory predicts that the phases in the grain lattice become coherent over macroscopic distances and that, therefore, the lattice exhibits zero resistivity to an infinitesimal externally applied electric field. Mathematically, it is the temperature at which equations 12-14 first exhibit a non-zero solution for the phase variables. The equation for T_c is readily obtained from equations 12-14 by noting that near T_c, the phase variables are all small, and so, therefore, are the expectation values appearing in equation 14. We may therefore expand the exponential in equation 12 as $1 - H_{eff}/k_B T$, keeping the first term since the zeroth order integral vanishes. Carrying out the phase integral, we then obtain a linearized mean-field equation,

$$\eta_i - \frac{1}{2k_B T} \sum_j J_{ij} \exp(iA_{ij})\eta_j = 0 \quad (15)$$

This is precisely the Schrodinger equation in the tight-binding representation for an "electron" (of charge 2e) in a magnetic field B and moving in a lattice of N sites. The order parameter η_i is the analog of the complex wave function at the ith site, J_{ij} is the hopping integral, and $\beta = (k_B T)^{-1}$ is the energy eigenvalue. T_c is the highest value of T for which equation 15 has a non-trivial solution, that is, it maps onto the band edge of the tight-binding electron problem. These band edges have been worked out by Hofstadter [15] for the two-dimensional square

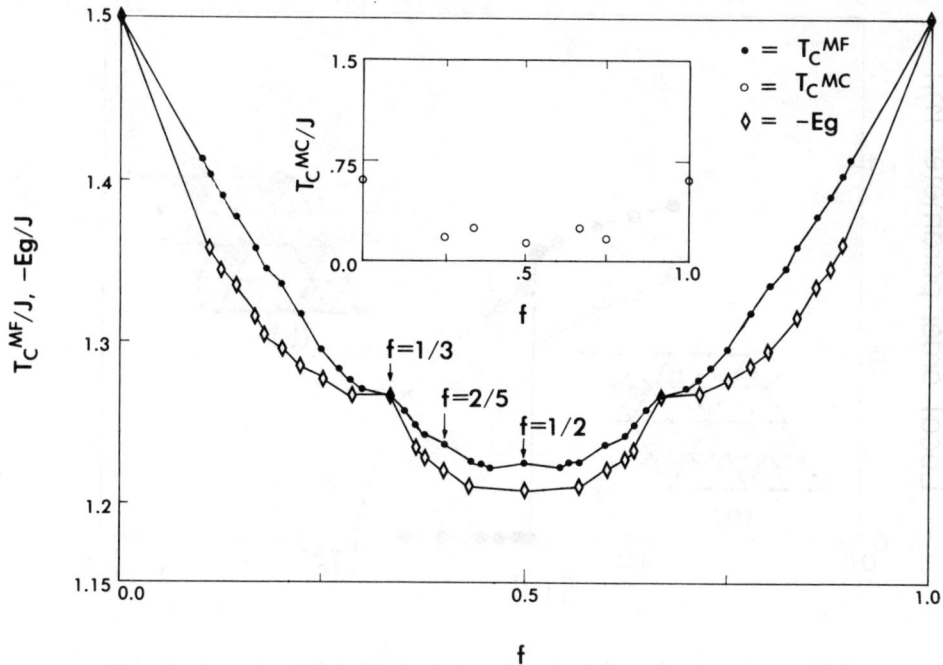

Figure 3. Same as Figure 2, except for honeycomb lattice. Ground state energies are denoted by diamonds, and f is the field in units of flux quanta per primitive hexagon. After W.Y. Shih, Ph.D. dissertation.

lattice, and by Claro and Wannier for the triangular lattice [16]. It may be verified that the tight-binding band edges do indeed map onto the mean-field transition temperatures for these lattices.

Figures 2 and 3 shows the mean-field transition temperatures for the triangular, and honeycomb lattices with nearest-neighbor overlaps. The most immediately striking feature of these curves is their irregularity: they have an amazing amount of structure. Closer inspection of these curves (i.e. solution for T_c^{MF} at f=p/q with p and q integer and relatively large values of q) shows that this irregularity persists on a very fine scale. Indeed, as was first pointed out by Hofstadter, the curve for $T_c^{MF}(f)$ (the band edge in his problem) shows self-similarity down to a very fine scale. That is, if certain small sections of the curve are blown up in scale, they closely resemble the entire curve.

The curves also show certain characteristic differences between the lattices. Although all three curves are periodic in f with period 1, the triangular lattice (as well as the square lattice which is not plotted here) shows a conspicuous secondary maximum at half-integer f (in addition to the primary maximum at integer f), while the secondary maximum at half-integer f is far weaker or non-existent for the honeycomb lattice. Evidence for this secondary maximum has been seen experimentally, as discussed below. The triangular lattice, besides the maximum at f=1/2,

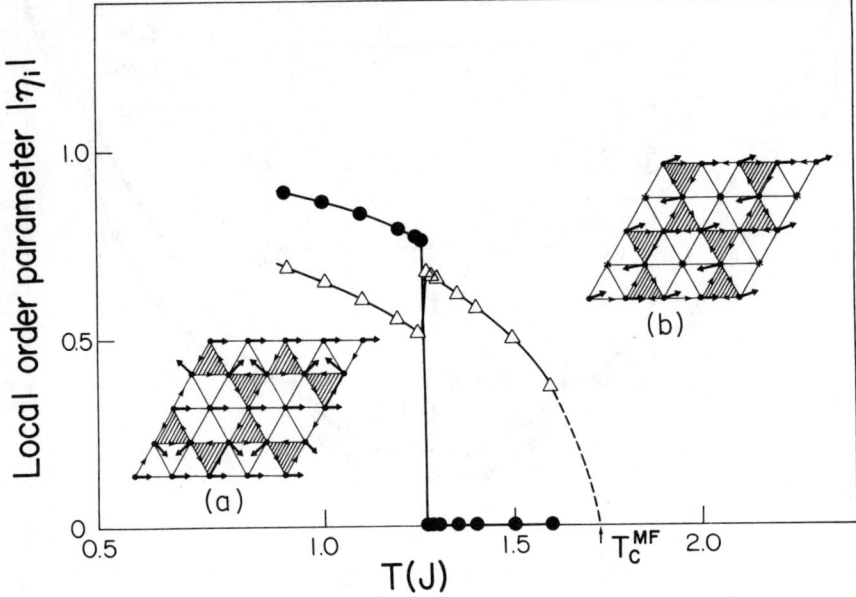

Figure 4. Mean-field local order parameter γ versus temperature T for f=1/4 in a triangular lattice. The abrupt drop of the order parameter at the center of the hexagon at the lower transition at T_{c1} = 1.27 J suggests that the lower transition is first order. The triangles denote the order parameter at the periphery of the hexagons. Inserts (a) and (b) show the phase configuration of the ground state and of the intermediate state. All local order parameters vanish at T_{c2} = 1.73 J. After W.Y. Shih, Ph.D. dissertation.

also shows striking structure at f=1/4 and 3/4, which is absent from the square lattice. This quarter-integer structure has also been seen experimentally (see below).

The transition at f=1/4 in the triangular lattice proves to be unusual in other respects as well. If the full mean-field equations are solved below T_c^{MF}, it is found that there are two phase transitions [17]. Just below the upper transition, at T_{c2}^{MF}, three-quarters of the η_i's become non-zero, while one-quarter remain zero. At the lower transition, which is first order within mean-field theory, all the phase order parameters change discontinuously, as illustrated in figure 4, and below this temperature (T_{c2}^{MF}) all are non-zero. In the intermediate temperature range, in this approximation, some of the superconducting phases remain random, or equivalently, some of the bonds are superconducting while some remain normal. In the analogous tight-binding picture, the wave function for the band-edge state at f=1/4 in the triangular lattice has non-zero amplitude at only 3/4 of the sites. It should be emphasized, however, that this double transition is clearly present only in molecular field theory. In

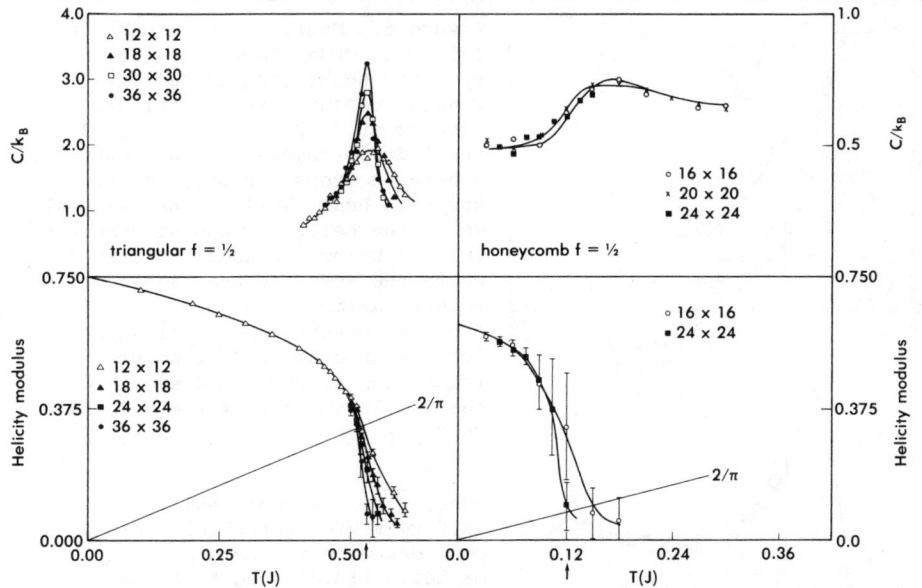

Figure 5. Specific heat C_V and helicity modulus, as calculated via Monte Carlo simulation at f=1/2 in the triangular and the honeycomb lattice. Different curves denote different Monte Carlo cell sizes. After W.Y. Shih, Ph.D. dissertation.

the Monte Carlo studies (see below) numerical evidence is quite ambiguous and the existence of this double transition cannot be said to have been proved at this stage.

The most convenient way to obtain results more accurate than mean-field theory is by computer experiment, i.e. by Monte Carlo simulation of the model Hamiltonian, equation 5 [18]. Since equation 5 is a classical Hamiltonian, it is very convenient to use such simulation, within standard algorithms, to find canonical expectation values of various quantities of interest. The transition temperatures resulting from such simulations are generally much lower than the corresponding mean-field values, as ilustrated in figures 2 and 3, but the relative order of the transition temperatures generally remains the same.

The MC simulations not only give $T_c(f)$, but also provide suggestive information about the universality classes of the various phase transitions. The clearest information is provided by plots of the specific heat and by the so-called helicity modulus as a function of temperature. The latter is the analog of the spin-wave stiffness in ferromagnets, and represents the resistance of the phases to an externally applied "twist". Such a twist is possible to apply experimentally (see below). Explicitly, one defines this stiffness by considering a d-dimensional superconducting array with periodic boundary conditions. The helicity modulus γ is a d x d matrix whose components are defined by [19]

$$\gamma_{ij} = \left(\frac{\partial^2 F}{\partial A'_i \partial A'_j}\right)_{\vec{A}' = 0} \qquad (16)$$

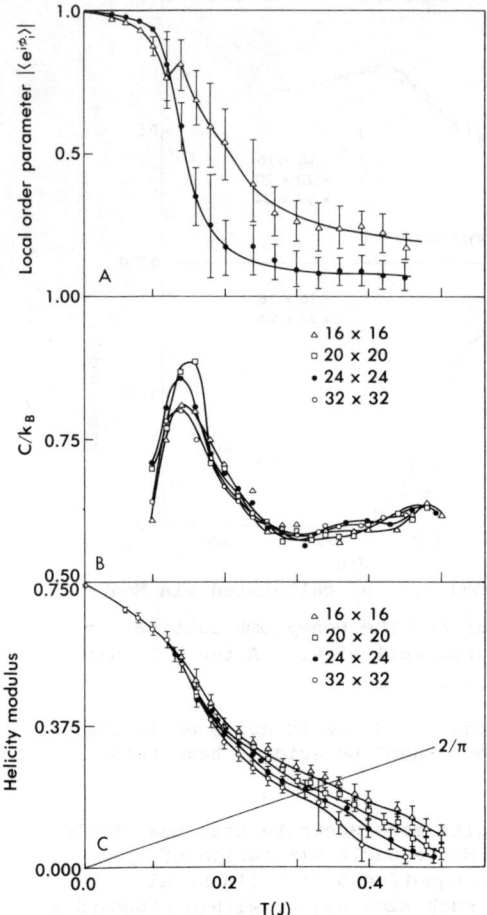

Figure 6. Monte Carlo results for (a) local order parameter, (b) specific heat, and (c) helicity modulus versus T in the triangular lattice at f=1/4. Around T=0.13 J, the order parameter at the center of a hexagon drops abruptly and the specific heat shows an unusual peak while the helicity modulus persists, falling below $2/\pi$ near T = 0.25 J, where the specific heat shows a weaker maximum. These results provide somewhat equivocal support for the hypothesis of a double transition as predicted by mean-field theory. After W.Y. Shih, Ph.D. dissertation.

where $\vec{A}'$ is an additional uniform vector potential (beyond the one connected to the applied magnetic field), and F is the Helmholtz free energy. If one uses the relation $F = -k_B T \ln Z$, where Z is the partition function, and carries out the indicated derivatives in equation 13, one finds that each element of the helicity modulus matrix can expressed in terms of canonical expectation values. The matrix can therefore be computed by standard MC techniques.

Figure 5 shows the specific heat and helicity modulus for two representative transitions in two-dimensional lattices - the triangular and the honeycomb lattices at f=1/2. (In both, γ is isotropic, i.e., a multiple of the unit tensor.) The transition in the triangular lattice at f=1/2 shows clear evidence of a singularity in C_V - the peak is strongly

dependent on Monte Carlo cell size, becoming progressively sharper as the cell grows. Similar results were obtained earlier by Teitel and Jayaprakash for the square lattice; they interpreted their results to indicate that f=1/2 for this lattice is a conventional second order phase transition - probably in the same universality class as the two-dimensional Ising model. In contrast, the honeycomb lattice at f=1/2 shows no size dependence in the peak of C_V, and hence no singularity. This result

suggests a continuous phase transition of a more unusual type - possibly in the universality class of the Kosterlitz-Thouless vortex-unbinding transition first proposed for the unfrustrated d=2 xy model. Transitions of this kind are characterized by a universal drop in γ at T_c, of magnitude $2k_B T_c/\Pi$. In order to see such a drop in the numerical data, one

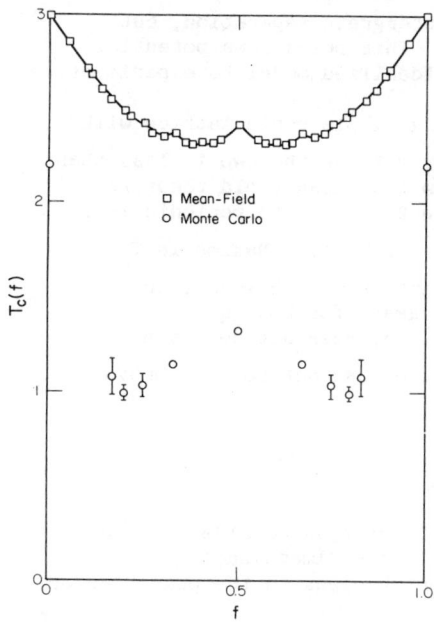

Figure 7. Mean-field (squares) and Monte Carlo (circles) transition temperatures for a simple cubic lattice as a function of f, for a field parallel to one of the axes of the lattice. After W.Y. Shih, Ph.D. dissertation.

constructs a straight line of slope $2/\Pi$ in a plot of $\gamma(T)$. If T_c is taken as the point of intersection of $\gamma(T)$ with this straight line, then $\gamma(T)$ at this temperature should show a discontinuous drop to zero. The presence of this drop is not clear in the honeycomb lattice, and thus it is not obvious that this transition is in the KT universality class.

Still, the transition is evidently of a different character than for the triangular lattice.

Figure 6 shows the Monte Carlo data for the triangular lattice at f=1/4, the field for which mean-field theory predicts a double transition. While γ and C_V both clearly behave differently from most other transitions in the lattices studied by MC (as exemplified by figure 5), the presence of two transitions is not unequivocally established by these data and more work remains to be done to complete the demonstration.

B. Three-dimensional lattices

Three-dimensional ordered lattices of superconducting grains in a non-superconducting host are clearly more difficult to prepare experimentally than are two-dimensional. Nonetheless, they are of some interest for purposes of comparison with the two-dimensional results, and for contrast with disordered arrays, which can easily be manufactured in both two and three dimensional form.

Arrays in three dimensions present several new qualitative features [19], most easily seen by considering simple cubic arrays. First, the transition temperature T_c is a function of the vector $\vec{f} = (f_x, f_y, f_z)$ (where f_i is defined as the flux through an elementary square perpendicular to the ith coordinate axis), not just the magnitude of the field. Both mean-field calculations and simulations confirm that T_c is indeed a highly anisotropic function. A second point is that the helicity modulus tensor γ_{ij} is anisotropic and not, as in most two-dimensional arrays, a multiple of the unit tensor. For a simple cubic array with field parallel to the z axis, for example, γ is uniaxial, with different stiffness constants parallel and perpendicular to the field. Finally, in contrast to two-dimensional arrays, one has in three-dimensional systems a more difficult problem in determining the local field. The penetration depth in

3d arrays may be many times larger than the intergrain separation, but still smaller than typical sample dimensions. This poses some potential difficulty in comparing predictions from our idealized model to experiment.

Figure 7 shows mean-field and MC T_c's for a simple cubic lattice with field parallel to the z axis. The discrepancy between the two is less than 2-d arrays, in agreement with the usual dictum that mean-field theories become better at higher dimensionality. As in 2d, $T_c(f)$ is periodic in f with a period of one flux quantum per square (i.e. f=1). Maxima in T_c occur at integer f, secondary maxima at half-integer f. However, in contrast to a speculation of Teitel and Jayaprakash for 2-d square lattices, numerical evidence suggests that $T_c(1/q)$ does not decrease monotonically with increasing q (this speculation has not been disproved for square lattices).

IV. Disordered Arrays

The physics of the Hamiltonian, equation 5, changes considerably in disordered arrays, whether two-dimensional or three-dimensional. This change is easily understood by considering what happens to the phase factor A_{ij} in a disordered sample.

Consider an array in which the positions $\vec{x}_i$ of the grains are random, in the sense that the vectors $\vec{x}_i - \vec{x}_j$ are not lattice vectors. Then the phase factors A_{ij} will also have a certain random distribution. For zero external magnetic field, all these factors will vanish, of course, but all will increase linearly with B, except (in the gauge being considered) for those bonds which are precisely oriented in the x-z plane. If $\bar{B}$ is sufficiently large, these phase factors will have a distribution with a width large compared to 2Π. Thus the interactions between grains will be distributed in preferred orientation: some will be ferromagnetic, some antiferromagnetic, and some will tend to orient the phases at angles other than zero or 180°. Since this distribution will be random, the array of grains will behave like an x-y or two-component spin glass [20]. More precisely, the array is a gauge glass similar to that first discussed by Hertz [20].

For zero field, the interactions between "spins" are all ferromagnetic, and the composite behaves like an x-y disordered ferromagnet. Thus, as the field increases, we expect a contiuous transformation from typical (although disordered) ferromagnetic behavior to spin-glass behavior. This transformation can be achieved by varying an early controlled experimental parameter: the magnetic field. The (gradual) crossover from one behavior to the other should occur at a field of about one flux quantum per typical closed loop of nearest neighbor bonds: $B_c = \phi_0/d^2$. For typical random distributions with intergrain separations of order 1 - 100 A, this field varies from 0.1 gauss to 10 k gauss.

A schematic of the expected behavior (in either two or three dimensions) is shown if figure 8 for (a) ordered arrays, (b) "weakly" disordered arrays and (c) "strongly" disordered arrays. The difference

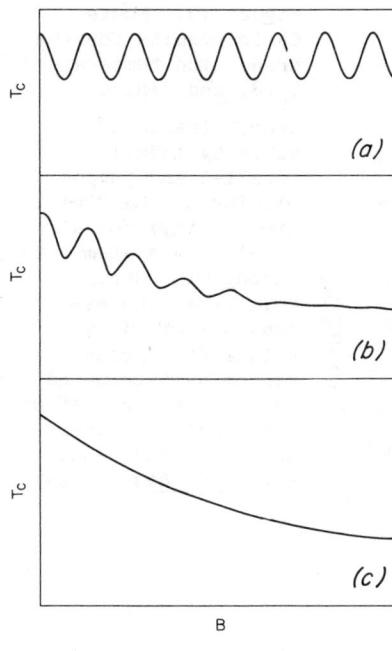

Figure 8. A schematic of the expected behavior of $T_c(B)$ for (a) an ordered array, (b) an array with weak disorder, (c) an array with strong disorder. After W.Y. Shih, Ph.D. dissertation.

between (b) and (c) is that, in the former, all closed loops have nearly the same projected areas perpendicular to the field, while in the latter, the loops have a much wider distribution of areas. Thus, in weakly disordered arrays, there exist certain non-zero fields such that the flux through any closed loop is nearly an integer multiple of a flux quantum, and the frustration is very small. At strong fields, the frustration eventually becomes nearly field independent, and the transition temperature should saturate, for both strong and weak disorder. The limiting "high-field" value of $T_c(B)$ is lower than $T_c(0)$ because frustration makes it more difficult for the phases to order. The schematic oscillations shown in Figure 8b reflect the minima in frustration at certain well-defined fields, for the weakly disordered samples. Such minima do not occur in strongly disordered samples, because the frustration in effect increases monotonically with field.

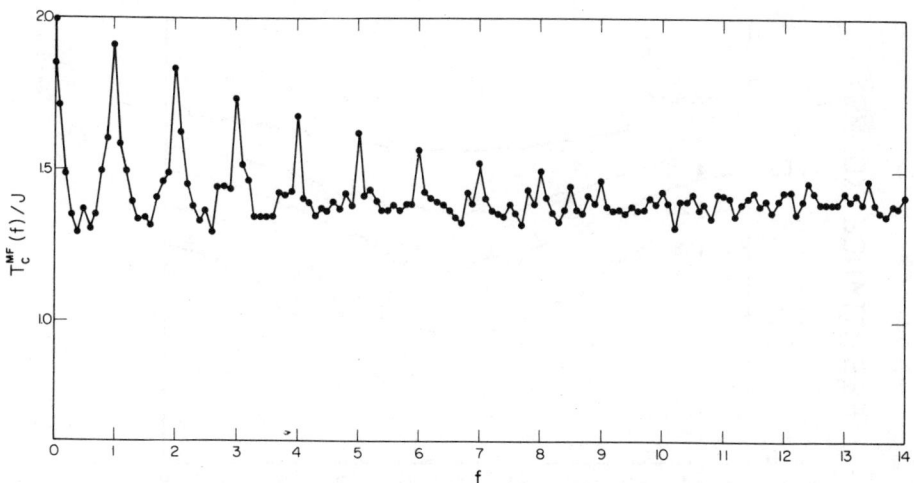

Figure 9. Mean-field transition temperature $T_c^{MF}(f)$ versus f for a weakly disordered two-dimensional square array. The disorder consists of random slight errors in positioning of the grains at their nominal lattice sites. Apparently random "noise" at high f is real structure in the mean-field calculation. After W.Y. Shih, Ph.D. dissertation.

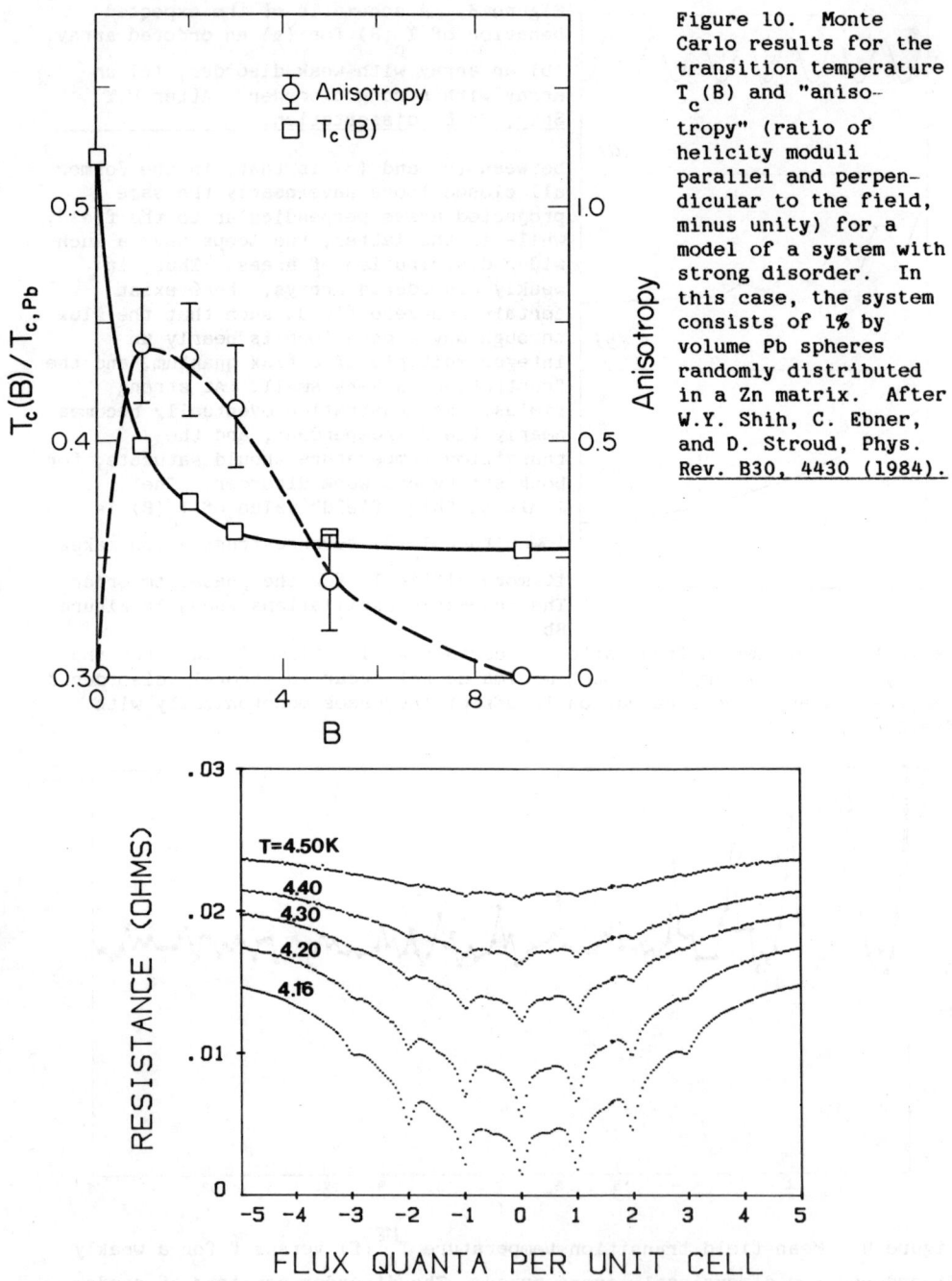

Figure 10. Monte Carlo results for the transition temperature $T_c(B)$ and "anisotropy" (ratio of helicity moduli parallel and perpendicular to the field, minus unity) for a model of a system with strong disorder. In this case, the system consists of 1% by volume Pb spheres randomly distributed in a Zn matrix. After W.Y. Shih, C. Ebner, and D. Stroud, Phys. Rev. B30, 4430 (1984).

Figure 11. Resistance versus magnetic field for the disk arrays of Figure 1, for several temperatures above $T_c(B)$. After R. Brown, D. Rudman, and J.C. Garland (to be published).

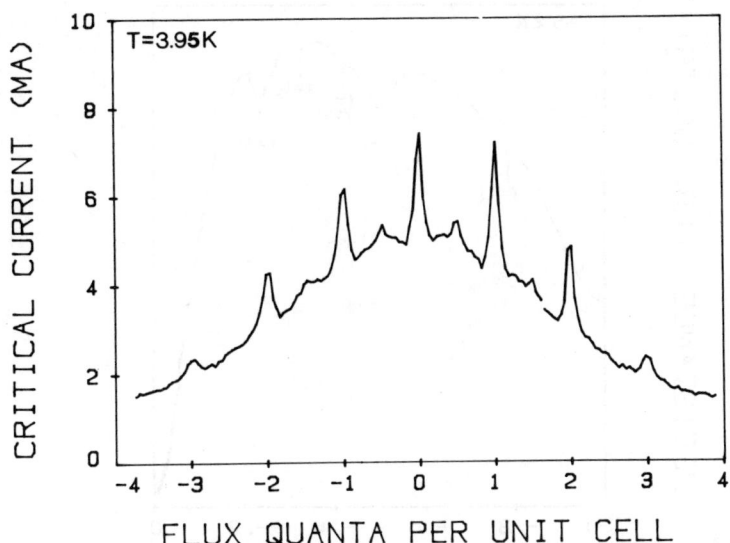

Figure 12. Critical current versus magnetic field for the disk arrays of Figure 1, for a temperature below $T_c(B)$. After R. Brown, D. Rudman, and J.C. Garland (to be published).

To illustrate the cases of weak and strong disorder, we show in figures 9 and 10 and 13 two numerical calculations. The weak-disorder illustration [18] is executed by mean-field theory, and represents the effects of slight errors in positioning the disks in a square array. The disks are assumed to deviate randomly in the x-y plane by up to 1% from their nominal centers, but the coupling constants at zero field themselves are assumed unaffected by these errors. The characteristic exonentially decaying envelope is superimposed on the principal and secondary maxima of $T_c(B)$ seen in the ordered array. Strong disorder [19] is shown by a model of a dilute suspension of Pb spheres in a Zn host. (Here Zn, a superconductor with a lower T_c than Pb, plays the role of the normal metal.) $T_c(B)$ in this case, as calculated by Monte Carlo simulation, is monotonically decreasing with field, and the characteristic cross-over field, for the typical experimental parameters, is about 10 g.

The spin glass regime in these disordered superconductors, like more conventional spin glasses, is subject to various uncertainties, most particularly in whether or not a true phase transition has actually taken place at $T_c(B)$, the glass transition temperature. Neither mean-field nor MC simulations can really answer these questions. They can only say that below a certain critical temperature, the system has frozen into some particular metastable configuration. There may still be occasional fluctuations out of this configuration, as in other spin glasses.

V. Comparison to Experiment

A variety of experiments have provided substantial confirmation of the results described here, especially for the ordered, two-dimensional

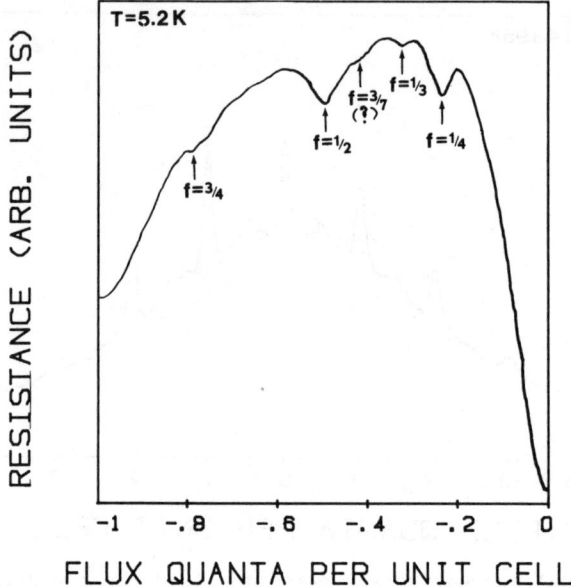

Figure 13. Resistance versus magnetic field for the asterisk arrays of Figure 1, for a temperature above $T_c(B)$. After R. Brown, D. Rudman, and J.C. Garland (to be published).

arrays. Experiments have been carried out on S-N-S arrays at several institutions [9, 10, 11, 21, 22], on S-I-S arrays (i.e. tunnel junctions) at IBM [18] and on the closely similar superconducting wire networks at Grenoble [4]. A number of results from the Ohio State experiments are displayed in figures 11 to 13 by way of illustration. Note that none of these experiments measure $T_c(B)$ directly. Rather, what is measured is the magnetoresistance above the transition temperature, or the critical current below. Both are expected to reflect the underlying periodicity of the $T_c(B)$ curves. Above $T_c(B)$, the magnetoresistance should have dips at fields where $T_c(B)$ has maxima, reflecting a lowered resistance resulting from being relatively close to the transition temperature. Below $T_c(B)$, the critical current $I_c(B)$ should have maxima at the corresponding field, because at these fields the system is relatively farthest from $T_c(B)$.

The data from the Ohio State group is based on a triangular lattice, and suggests a transition temperature $T_c(B)$ curve which (i) is periodic in B with a period of one flux quantum per elementary triangle, (ii) has a maximum at integral values of f, (iii) has secondary maxima at half-integral values of f, (iv) has tertiary maxima at quarter-integral values, (v) has an envelope which slowly decays with increasing field. Predictions (i) - (iv) are in detailed agreement with the mean-field theory figure 2, even though the mean-field approximation is expected to fail in two-dimensions. [There is no experimental evidence for or against the predicted second transition at f=1/4, nor is it clear just what the experimental signature of such a transition would be.] The success of mean-field theory is not unique: the Monte Carlo simulations also give principal, secondary, and tertiary maxima in the same order, though at much lower temperatures. Taken as a whole, however, these results give rather convincing evidence that the model Hamiltonian, equation 5, used to describe these experiments includes much of the correct physics. As for the decreasing envelope, this could be evidence of the weakly disordered "spin-glass" discussed above, but more likely it is simply a result of a field-dependent interaction, $J = J(B)$, which could arise from the non-point-like nature of the interactions.

Confrontation between theory and experiment is far from complete at this writing. The theory so far is able to predict only the transition temperatures and ground state energies, whereas the experiments measure transport properties such as resistivity and critical current [22]. These are more difficult to compute, and work to calculate them explicitly is much needed. Comparison between theory and experiment for three-dimensional systems, and disordered systems, is lacking, in part because of the great difficulty in making ordered arrays in three dimensions or of truly controlling the disorder. Again, theory at this stage focuses mostly on quantities which are not directly measured experimentally - a focus that it would be desirable to alter [23]. It is to be hoped that a more thorough comparison will be available in the next year or two.

ACKNOWLEDGEMENTS

The authors are most grateful to Professor J.C. Garland for valuable discussions, and for permission to use date from his group in advance of publication. They also thank Professor C. Jayaprakash and Dr. S. Teitel for many enlightening conversations.

REFERENCES

1) For many review of work prior to 1979, see the articles in Gubser, D.U., et. al., eds.: Inhomogeneous Superconductors - 1979, AIP Conf. Proc. 58 (American Institute of Physics, New York, 1980).
2) For more recent reviews, see Goldman, A.M. and Wolf, S.A., eds.: Percolation, Localization and Superconductivity (Plenum Press, New York and London, 1984).
3) Muhlschlegel, B., in ref. 2, pp. 65-82.
4) Pannetier, B., Chaussy, J. and Rammal, R.: J. Phys. (France) (1983), L853.
5) Kosterlitz, J.M. and Thouless, D.J.: J. Phys. (1973) C7, 1181.
6) Berezinskii, V.L.: Zh. Eksp. Theor. Fiz. (1971) 61, 1144 [Sov. Phys. JETP (1972) 34, 610]. For reviews of recent theory and experiment, see the articles by Minnhagen and Mooij in Reference 2).
7) Anderson, P.W., in Caianello, E.R., ed: Lectures on the Many-Body Problem (Academic Press, New York 1964), Vol. 2, p. 127.
8) Webb, R.A., Voss, R.F., Grinstein, G. and Horn, P.M.: Phys. Rev. Lett. (1983) 51, 690.
9) Kimhi, D., Leyvraz, F. and Ariosa, D.: Phys. Rev. (1984) B29, 1487.
10) Resnick, D.J. Garland, J.C., Boyd, J.T., Shoemaker, S. and Newrock, R.S.: Phys. Rev. Lett. (1981) 47, 1542.

11) Tinkham, M., Abraham, D.W. and Lobb, C.J.: Phys. Rev. (1983) B28, 6578; Lobb, C.J., Abraham, D.W. and Tinkham, M.: Phys. Rev. (1983) B27, 150.
12) Teitel, S. and Jayaprakash, C.: Phys. Rev. (1983), B27, 598; Teitel, S. and Jayaprakash, C.: Phys. Rev. Lett. (1983) 51, 1999.
13) Toulouse, G.: Commun. Phys. (1977) 2, 155.
14) Shih, W.Y. and Stroud, D.: Phys. Rev. (1983) B28, 6575.
15) Hofstadter, D.R.: Phys. Rev. (1976) B14, 2239.
16) Claro, F.H. and Wannier, G.H.: Phys Rev. (1979) B19, 6068.
17) Shih, W.Y. and Stroud, D.: Phys. Rev. (in press).
18) Shih, W.Y.: Ph.D. Dissertation (Ohio State University, 1984) (unpublished).
19) Shih, W.Y., Ebner, C. and Stroud, D.: Phys. Rev. (1984) B30, 134.
20) Hertz, J.A. Phys. Rev. (1978) A18, 4875.
21) Brown, R. L., Rudman, D. and Garland, J.C., (unpublished).
22) Resnick, D.J., Brown, R.K., Rudman, D.A., Garland, J.C. and Newrock, R.S.: Proceedings of the 17th International Conference on Low Temperature Physics LT-17 (North-Holland, Amsterdam, 1984), pp. 739-40.
23) The same point has been made by Lobb, C.J.: Proceedings of the International Conference on Low Temperature Physics, LT-17 (North-Holland, Amsterdam, 1984).

Note added in proof: M.Y. Choi and S. Doniach (preprint) have recently found, using a Ginzburg-Landau-Wilson expansion, that the phase ordering transition in the triangular lattice at $f=1/2$ is in the same universality class as the three-state Potts model. It is thus a continuous transition, as is implied by the simulations of W.Y. Shih. Thomas Halsey (preprints) has calculated the ground state energy and critical current for the square lattice as a function of f, using certain assumptions about the vortex configuration. S. Teitel and C. Jayaprakash have extended their earlier studies to superconducting wire networks in a recent preprint.

This work has been supported in part by NSF through grant DMR 81-14842 and through the Materials Research Laboratory at Ohio State University, grant DMR 81-19368.

ELECTRONIC STRUCTURE OF ALLOY SEMICONDUCTORS

L. C. Davis
Research Staff
Ford Motor Company, Dearborn, MI 48121-2053

ABSTRACT

The Haydock recursion method is used to calculate the electronic structure of substitutional alloys. Exact results for a random, linear chain and a three-dimensional model alloy are compared to the recursion method. Spectral weight functions from the coherent-potential approximation for $Hg_{1-x}Cd_xTe$ agree well with those calculated in the present work. Two recent models for $(GaAs)_{1-x}Ge_{2x}$ are analyzed. The existence of an energy gap has been found to depend sensitively on short-range order.

INTRODUCTION

Alloy semiconductors are technologically useful materials, especially for electro-optical devices. The variation of the optical absorption edge with alloy composition can be used to tailor device properties. Understanding this fundamental property is an important and challenging task.

For the familiar III-V alloys (e.g., $Ga_{1-x}Al_xAs$), a composition-dependent band structure derived from the virtual-crystal approximation (VCA) provides a good description [1,2]. Although fundamentally incorrect because translational symmetry is lacking, the VCA is adequate because the potential fluctuations are small compared to the band widths. The effects of disorder are not so large as to invalidate the band picture.

In this paper semiconductor alloys for which the VCA is inappropriate will be discussed. To go beyond VCA, typically the coherent-potential approximation (CPA) [1,2] is used. This method is computationally tractable and has been used extensively in alloy calculations, but it is not clear to what extent the results are reliable and how systematically to improve the approximation. In addition, for some alloys short-range order is important and this is not straightforward to introduce into the CPA.

The Haydock recursion method [3] has proven to be an extremely powerful technique for the calculation of the density of states (DOS) of disordered materials. Recently, I have shown that spectral weights (projection of the eigenfunctions on Bloch states) can be calculated by this method as well [4]. In many respects, these functions are more useful than DOS. The purpose of this paper is to calculate the effects of substitutional disorder and short-range order on the electronic structure of semiconductor alloys with the Haydock recursion method.

Formal Theory

In this section, the formal theory for the calculation of the spectral weight function is briefly presented. A more detailed description can be found in Ref. [4]. We assume that the alloy can be described by a tight-binding approximation with an atomic-like, orthonormal basis set at each site. Typically, nearest- and next-nearest-neighbor interactions are included. A large ($\sim$1600 atoms) cluster of the alloy is generated with either random order or with some specified short-range order. The spectral weight or the DOS is then calculated using the Haydock recursion method. It is important to realize that in the limit of infinitely large clusters and infinite number of recursions, the exact result (within tight-binding) is approached. Generally, periodic boundary conditions are imposed.

One chooses a start state u_0 according to the type of projected DOS of interest, e.g., a Bloch state in the case of a spectral weight function or a p state in the case of a p-partial DOS. The Hamiltonian is transformed to tridiagonal form (chain model) [3] by solving the three-term recursion

$$Hu_n = a_n u_n + b_n u_{n-1} + b_{n+1} u_{n+1}, \qquad (1)$$

where the u_n are orthonormal, $u_{-1} = 0$,

$$(u_n|H|u_n) = a_n \qquad (2)$$

and

$$(u_{n+1}|H|u_n) = b_{n+1}. \qquad (3)$$

The projected DOS is defined by

$$n_0(E) = \sum_m (u_0|\Psi_m)^2 \delta(E - E_m), \qquad (4)$$

where Ψ_m and E_m are the eigenfunctions and eigenvalues of H. These are not actually calculated, instead $n_0(E)$ is written as the imaginary part (divided by $-\pi$) of a Green's function that can be expanded in a continued fraction [3,4] involving the recursion coefficients a_n and b_n. In my numerical work I use subroutines from the Cambridge Recursion Library [5], in which the integrated DOS

$$N_0(E) = \int_{-\infty}^{E} n_0(E') \, dE' \qquad (5)$$

is computed from the recursion coefficients. It can be shown [3] that rigorous upper and lower bounds to $N_0(E)$ can be determined (see Fig. 1). Currently, the average of these bounds is used and then numerically differentiated to obtain $n_0(E)$.

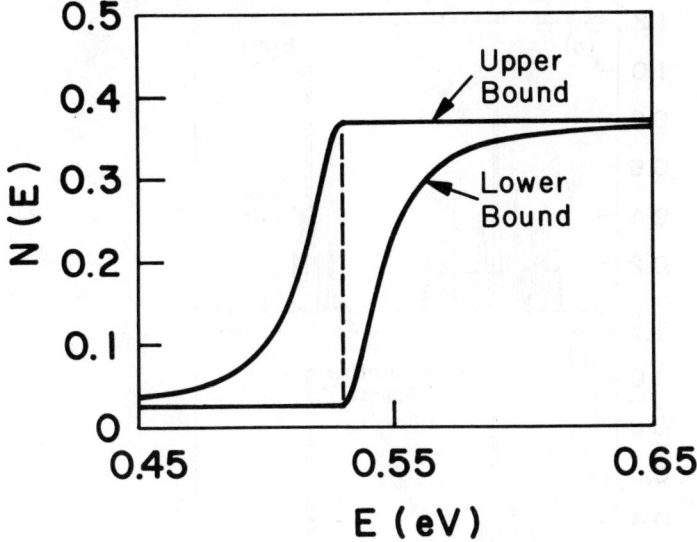

Figure 1. Upper and lower bounds on the integrated density of states, N(E), near an isolated eigenvalue (dashed line). The exact result is a step function at the eigenvalue.

Comparison to exact Results and the CPA

To demonstrate the accuracy of the recursion method, results from two exact model calculations are compared to the recursion method in this section. Also, CPA calculations for a realistic model of $Hg_{1-x}Cd_xTe$ [6] are compared for a point in the Brillouin zone where the VCA breaks down.

Figure 2 displays the DOS for a one-dimensional, random chain. The histogram represents exact results from Kaplan et al. [7] which are compared to single-site CPA in (a). A better approximation, the nearest-neighbor pair approximation, is shown in (b). The recursion method, (c), reproduces the fine structure much better. This is perhaps not so surprising since one linear chain has been transformed into another by the recursion method.

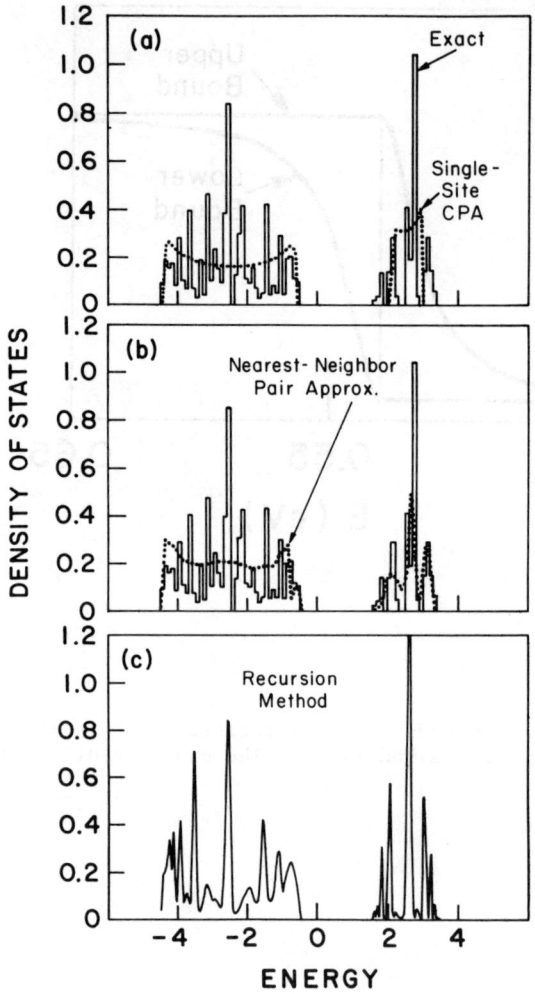

Figure 2. Density of states for a random, linear chain ($E_A = -E_B = -2.5$, $W_{AA} = 1$, $W_{AB} = W_{BB} = 0.5$, $c_B = 0.3$, see [7].) (a) Single-site coherent-potential approximation (CPA) compared to exact results. (b) Nearest-neighbor pair approximation (an improved CPA) compared to exact. (c) Recursion-method results for a 1000-atom chain (48 recursions, average of two runs).

A more severe test is the three-dimensional model of Fig. 3. Equation-of-motion techniques were used by Alben et al. [8] to compute the exact results. Again we see that the recursion method obtains more of the fine structure than the CPA does. The peak energies agree well for the majority band.

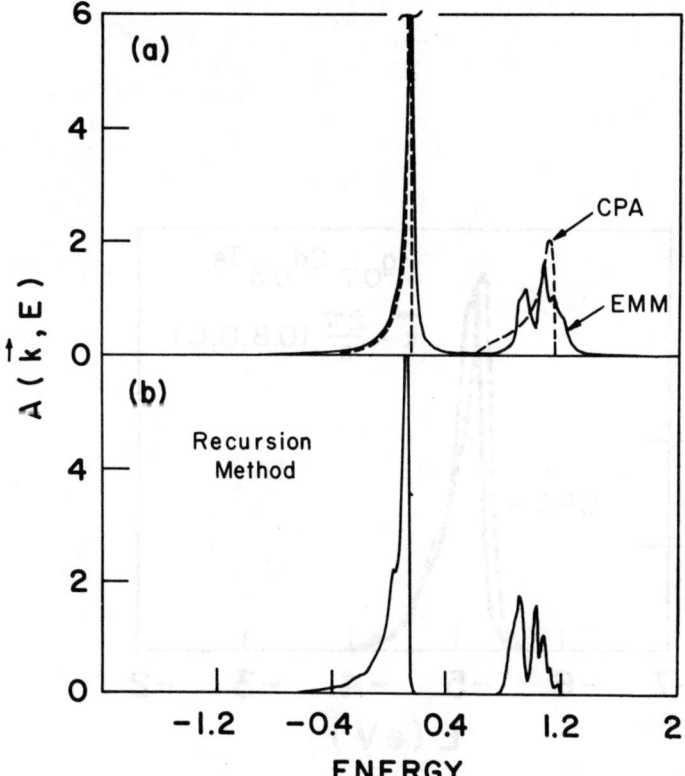

Figure 3. Spectral weight function $A(\vec{k},E)$ at $\vec{k} = 0$ for a three-dimensional alloy with diagonal disorder ($\delta = 1.5$, $c = 0.1$, simple cubic lattice, see [8].) (a) Single-site CPA and equation-of-motion (EMM) calculations. (b) Recursion method for a 16 × 14 × 12 cluster (48 recursions, average of five runs).

Hass et al. [6] have analyzed a model of $Hg_{1-x}Cd_xTe$ consisting of s and p functions on each site with nearest- and next-nearest neighbor interactions. Only site-diagonal disorder was retained (mostly in the large differences in cation s-orbital energies) and coupled, CPA self-consistent equations were solved. The energies of the peaks in the spectral weight functions agreed fairly well with the VCA, except for some deep-lying valence-band states near X and L. In Fig. 4, the CPA results are seen to be very close to those of the recursion method for a typical $\underline{k}$ point where the VCA fails. Since there is no fine structure in this example, CPA is expected to be good and the close agreement with the recursion method is reasonable. Comparable agreement was found at other points in the Brillouin zone, including those where VCA works well.

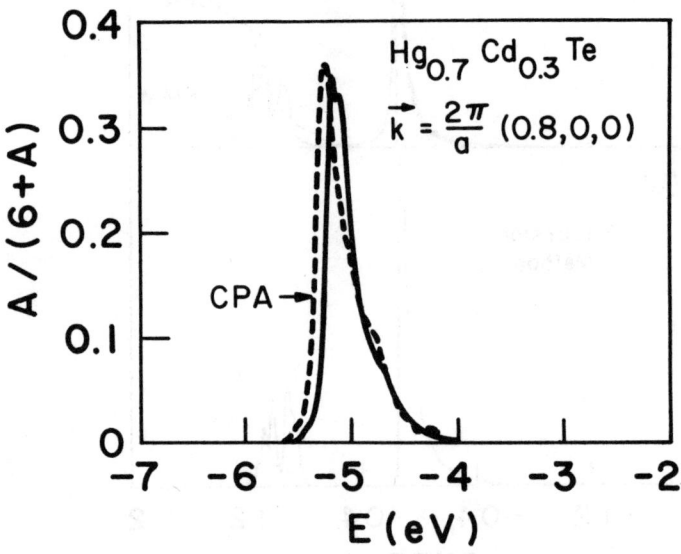

Figure 4. Spectral weight function for a realistic model of $Hg_{1-x}Cd_xTe$ using CPA [6] and recursion method (solid line). Omission of spin-orbit interaction in latter shifts peak upward slightly (10 x 12 x 14 cluster, 48 recursions, average of five runs).

Metastable $(GaAs)_{1-x}Ge_{2x}$ Alloys

Metastable alloys of GaAs and Ge have been prepared by sputtering [9] and by pyrolytic reactions [10]. The optical absorption edge (energy gap) shows deep bowing as a function of Ge concentration, x. Primarily on the basis of a possible kink in the curve of gap vs x, Newman and Dow [11] have suggested that a zinc-blende—diamond order-disorder transition occurs at x = 0.3. The crystal structure was analyzed by a mean-field theory for a Hamiltonian which is equivalent to that of a three-component spin model. The energy gap was determined from the mean-field order parameter and the VCA, modified to account for the predicted high concentration of As and Ga antisites.

Substitution of an As atom for a Ga perturbs the local potential substantially. In the tight-binding approximation, the p-orbital energies differ by 2.6 eV and the s by 5.7 eV whereas the GaAs energy gap is only 1.5 eV. Typical band widths are no greater than 7 eV, so the perturbation is as large as any other relevant energy and the VCA is apt to be inaccurate.

For x = 0.5, the order parameter in the Newman-Dow model is zero which implies that the probability of finding a Ge atom at any site is 0.5 and is 0.25 for each Ga and As. In mean-field theory there is no correlation between sites, i.e., no short-range order. It is straightforward to construct a random cluster with these statistics and to calculate spectral weights using the recursion method. The results for the tight-binding model of Ref. [11] are shown in Fig. 5 for $\underline{k}$ = 0. Γ_1 is the bottom of the conduction band and Γ_{15} the top of the valence band in GaAs. These states also comprise the direct gap in Ge. The VCA eigenvalue is close to the peak in the Γ_{15} function, but the value for Γ_1 is 0.6 eV higher. Consequently, we see that when the effects of disorder are accounted for, the energy gap vanishes for this model in sharp contrast to the VCA prediction. Also shown in Fig. 5 are the spectral weight functions for a cluster in which antisites are forbidden (order parameter equal 1-x). Ga (As) and Ge are equally probable on any site on the cation (anion) sublattice. For this cluster, the Γ_1 and Γ_{15} peaks do not overlap and are considerably narrower. Taking the separation of the peaks as the energy gap, we find a value much smaller than that from VCA but reasonably close to experiment.

It has been demonstrated that the elimination of As-As nearest-neighbor bonds is essential in obtaining the correct energy gap [12]. Physically this arises because the As antisite in GaAs has a deep bound state in the gap. A high concentration of As-As bonds therefore gives an impurity band which fills the gap. Clearly, without short-range order that forbids incorrect bonds (going beyond mean-field theory), the model of Newman and Dow is inconsistent with the experimental data.

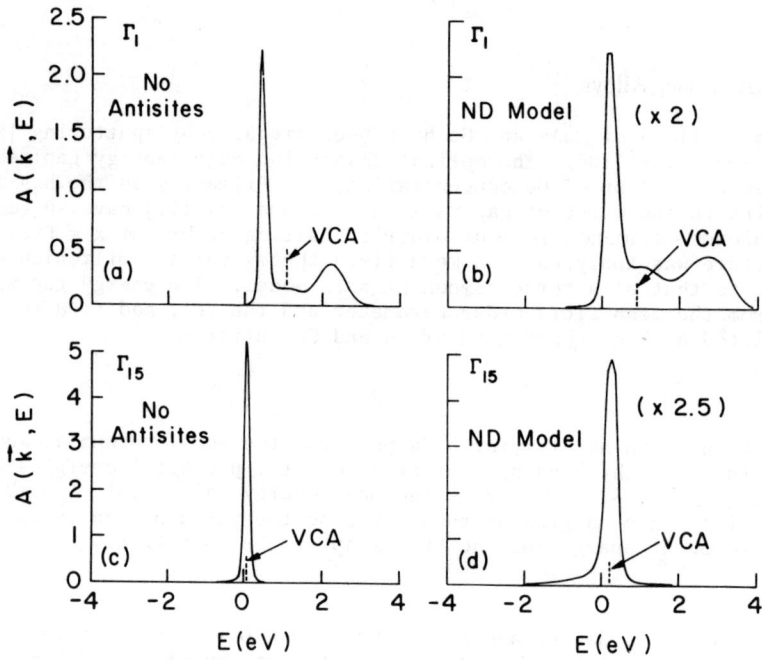

Figure 5. Recursion-method calculation of spectral weight functions at $\bar{k} = 0$ for $(GaAs)_{0.5}Ge$. Γ_1 is the bottom of the conduction band and Γ_{15} is the top of the valence band in zinc-blende notation. Units on ordinate are states/eV/spin/atom. In (a) and (c), the cluster was chosen randomly with probabilities $p(Ge) = 0.5$ and on the cation (anion) sublattice $p(Ga) = 0.5$ ($p(As) = 0.5$). In (b) and (d), at each site $p(Ge) = 0.5$ and $p(Ga) = p(As) = 0.25$ in accord with the mean-field description of Newman and Dow [11] (no correlations between sites). Eigenenergies in the virtual-crystal approximation (VCA) are indicated. The tight-binding model of [11] was used with no relative shift of the GaAs and Ge bands, i.e., both Γ_{15} energies are zero. (10 x 12 x 14 cluster, 48 recursions, average of five runs.)

Holloway and Davis [12] have presented a different structural model which is essentially an algorithm for generating a cluster. Three rules were postulated: (1) Ga and As atoms are always paired, (2) no Ga-Ga or As-As bonds occur, and (3) Ge and Ga-As pairs are otherwise randomly arranged. Spectral weights are shown in Fig. 6. Note that an (nonzero) offset between the Ge and GaAs bands has been included in this calculation. It shifts both peaks up, but has no other significant effect. The spectra closely resemble those of Fig. 5 for which antisites were forbidden.

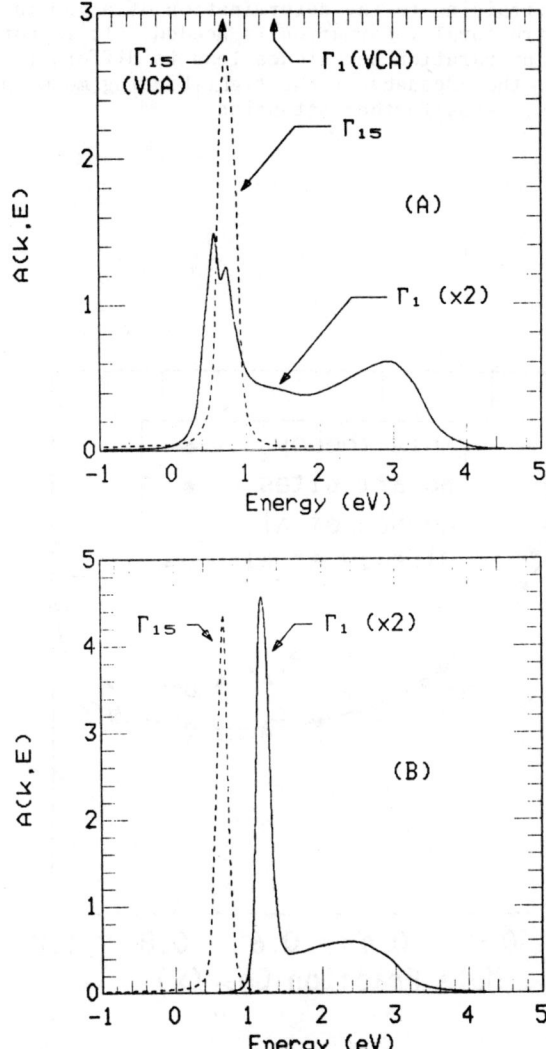

Figure 6. Spectral weight functions for $(GaAs)_{0.5}Ge$. (A) Same parameters as Figures 5b and 5d except for 1-eV shift upward of Ge bands. (B) Structural model of Holloway and Davis [12]. The principal difference is the absence of As-As bonds in the latter.

The curve of energy gap (separation between spectral-weight peaks) as a function of x for the Holloway-Davis model is displayed in Fig. 7. The agreement with experiment is quite good. However, there remain uncertainties in the experimental data pertaining to the determination of a gap in a disordered material. Also, more structural information is needed. It is not clear whether or not different preparation techniques lead to different structures. On the theoretical side, the adequacy of the tight-binding model, particularly for the conduction bands, needs further attention.

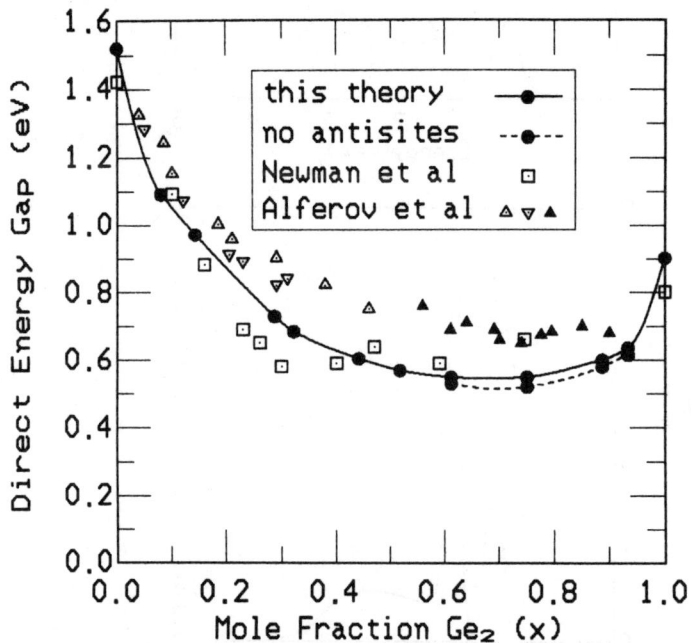

Figure 7. Energy gap for $(GaAs)_{1-x}Ge_{2x}$ (direct except near x = 1). Theoretical gap equals the energy difference between the Γ_1 and Γ_{15} spectral-weight peaks for the model of [12]. Experimental gaps from absorption measurements of [9] and [10]. Dashed curve was obtained by eliminating the antiphase disorder in the recursion-method clusters.

ACKNOWLEDGEMENTS

I would like to thank Drs. J. D. Dow, K. E. Newman, and K. C. Hass for useful discussions and correspondence. Dr. H. Holloway collaborated with me on the calculations reported in the last section

REFERENCES

1) Ehrenreich, H. and Schwartz, L. M. in Solid State Physics, edited by H. Ehrenreich, F. Seitz, and D. Turnbull (Academic, New York, 1976), Vol. 31, p. 149.
2) Chen, A.-B. and Sher, A.: Phys. Rev. (1981) B23, 5360.
3) Haydock, R. in Solid State Physics, edited by H. Ehrenreich, F. Seitz, and D. Turnbull (Academic, New York, 1980), Vol. 35, p. 215.
4) Davis, L. C.: Phys. Rev. (1983) B28, 6961.
5) Nex, C. M. M.: Cambridge Recursion Library (unpublished) (see Ref. [3], p. 78).
6) Hass, K. C., Ehrenreich, H., and Velický, B.: Phys. Rev. (1983) B27, 1088.
7) Kaplan, T., Leath, P. L., Gray, L. J., and Diehl, H. W.: Phys. Rev. (1980) B21, 4230.
8) Alben, R., Blume, M., Krakauer, H., and Schwartz, L.: Phys. Rev. (1975) B12, 4090.
9) Barnett, S. A., Ray, M. A., Lastras, A., Kramer, B., Greene, J. E., Raccah, P. M., and Abels, L. L.: Electron. Lett. (1982) 18, 891; Newman, K. E., Lastras-Martinez, A., Kramer, B., Barnett, S. A., Ray, M. A., Dow, J. D., Greene, J. E., and Raccah, P. M.: Phys. Rev. Lett. (1983) 50, 1466.
10) Alferov, Zh. I., Zhingarev, M. Z., Konnikov, S. G., Mokan, I. I., Ulin, V. P., Umanskii, V. E., and Yavich, B. S.: Sov. Phys. Semicond. (1982) 16, 532.
11) Newman, K. E. and Dow, J. D.: Phys. Rev. (1983) B27, 7495.
12) Holloway, H. and Davis, L. C.: Phys. Rev. Lett. (1984) 53, 830.

INDEX

INDEX

INDEX

Adhesive energy, 22
Adsorbate-adsorbate interactions, 1, 4
Adsorbed films, 1
anion-site bulk antisite defect, 42
anion on the cation site defect, 45
Anisotropy, crystalline, 73, 76
Argon, 3, 5, 14
Ag, 1, 10, 13, 37
Al, 21
    $Al_xGa_{1-x}As/Al$ contacts, 39
Amorphous semiconductors, 149
Amplitude fluctuation length, 145
Anomalous diffusion, 132
Antisite defects, 40
    bulk, 42
    four atom model, 41
    surface, 43
As,
    $Al_xGa_{1-x}As$, 197
    in contact with metals, 39
    $(GaAs)_{1-x}Ge_{2x}$, 197, 203
Au
    $Al_xGa_{1-x}As/Au$ contacts, 39
Atactic polymers, 107
Anharmonic effects, 6, 11
Antiferroelectric polymer, 109
anion species, 40
BCSOS model, 52
Biexciton, in semiconductors, 166, 172
Bilayer, 2, 3, 9
    Bilayer-to-trilayer transition, 15
Bimetallic interface, 21
Body-centered solid on solid model, 52
Bond angle field, 55
Boundary layer model, 73, 78
    steady state solutions, 85
Bound level, relation to localization, 150
Cantor sets, 131
Capillary,
    correction to Ivantsov model, 82
    waves, 53
cation species, 40
Cd,
    $Hg_{1-x}Cd_xTe$, 199, 202
Circular crystal interfaces, 81
Chaotic states, 119
Charge transfer, 27
Charging energy, in superconducting grains, 181
Chemical potential, 2, 3, 9, 15
Chemical reactions, autocatalytic, 74
Coherence length,
    connectivity, 133
    phase, 145
    superconducting, 179

Coherent potential approximation, 149, 197
Cohesive energy, 2
Commensurate phase, 119
Compressed monolayers, 4, 13
Compressibility, 2, 5, 15
Computer simulations, 3, 62-64, 101
Condensation, 3
Connectivity coherence length, 132
Contact potential, 25
Core energy, 55, 57, 65
Corrugated substrate surfaces, 9, 99
Correlation effects, 157
Correlation functions,
    density, in disordered electron systems, 149
    force-force, in diffusion on a fractal, 129, 136
    in the Laplacian roughening model, 57
Coulomb gas, two dimensions, 51, 99
Covalent network glasses, 127
Critical field, 178
Crystal growth, 1, 73
Crystal shapes, equilibrium, 69
Crystalline anisotropy, 73-76
Cu, 2, 10
    $CuBr$, 173
    $CuCl$, 173
Curvature, 74
dangling bonds, 40
    in anti-site defects, 4
De Boer parameter, 6
Debye-Waller correlation functions, 54
Defect,
    crystal, in curved three dimensional space, 93, 95
    levels, 39
    point, in solids, 155
Dendritic solidification, 73
Density of states, tailing, 152
Devonshire cell theory, 6
Diffusion,
    analogy to Schroedinger equation, 166
    field, thermal, 73
    limited aggregation, 77, 129
    on fractals, 129, 132
Dihedral angle forces, 125
Dimension,
    fractal, 145, 148
    of superconducting grains, 179
Dipoles, adsorption induced, 4
Discrete Gaussian model, 52, 65
Disclinations, 52
    core energy of, 55, 57, 65
    unbinding transition, 61

Dislocations, 52
  core energy, 55
Disorder,
  weak and strong, in superconducting arrays, 193
Dispersion force, 1
  London-van der Waals, 4
Dissipative systems, 74
Double sine-Gordon equation, 112
Droplet model, 130
Effective medium theories, 126
Einstein relation, 135
Elastic constants (at two dimensional melting transitions), 54
Elastic properties of random networks, 125
Electron gas, 166
Electron glass, classical, 99
Electron-hole complexes, 165
Electronic wave functions in disordered systems, 145
Equation of motion method, 201
Equilibrium crystal shapes, 69
Exchange-correlation energy, 21, 35
Exciton, in semiconductors, 165
Extended states, in disordered electron systems, 145
Fermi level pinning, 39
Ferroelectric crystalline polymer, 109
Field theories, of disordered electron systems, 148
Flame fronts, 74
Fluctuations,
  in anomalous diffusion, 135
  in disordered electron systems, 145
  length, 147
Folds, in Laplacian roughening model, 65
Force-force correlation, for diffusion on fractals, 129, 136
Fractals, 129
  dimensionality, 145, 148
  regular, 131
Fracton dimension, 135
Frank-van der Merwe mode, 2, 9, 11
Free boundary problem, 75, 77
Frenkel-Kontorova,
  map, 119
  model, 9
Friction, 22
Frustrated interactions, 9
Frustration, 93, 182
Ga,
  $Al_xGa_{1-x}As$ in contact with metals, 39
  $(GaAs)_{1-x}Ge_{2x}$, 199, 205
Gauge,
  choice in superconducting arrays, 181

glass, 190
Ge, 174
  $(GaAs)_{1-x}Ge_{2x}$, 197, 203
  Si_xGe_{1-x}, 38
Gelation, 130
Geometrical model, 75
Glassy phases, 93
Grain boundary energetics, 22
Graphite, 2, 3, 12, 14
Green's Function Monte Carlo Method, 165
  algorithm, 170
  importance sampling, 170
Hartree Fock technique, 155
Haydock recursion method, 197
He, 2, 13
  liquid and solid ^4He, 166
  Liquid ^3He, 166
  scattering from Pd(110), 10
Helicity modulus, in superconducting arrays, 187
Herringbone pattern, 8
Hexatic phase, 54
Hg,
  $Hg_{1-x}Cd_xTe$, 197, 199, 202
Holding potential, 2, 9, 15
Hydrodynamics instabilities, 74
Hyper-universal power law for diffusion on fractals, 130
Icosahedral,
  crystal, 94
  symmetry, 93
In,
  $Al_xGa_{1-x}As/In$ contacts, 35
Inert gases, 1
Inhomogeneous superconductors, 178
Interfacial kinetics, 73
Interfacial phase diagrams, 69
Irreversible aggregation, 130
Jellium, 23
Josephson tunneling, 177, 180
Koch curve, 131
Kossel crystals, 69
Kadanoff block-spin transformation, 62
Kinetic energy, 21, 35
Kinetics,
  interfacial, 73
  of molecular attachment, 75, 76
Kink, solutions to double sine Gordon equation, 112
Kohn-Sham equations, 23
Kosterlitz-Thouless transition, 54, 70, 181, 188
Kr, 2, 10, 14
Langevin equation, for anomalous dif-

fusion, 135
Laplacian roughening model, 51
  correlation functions, 57
  duality transformation, 55
  folds, 65
  in one dimension, 59
  Monte Carlo simulation, 60
  renormalization group, 62
Latent heat, 75
Lattice gas models, 51-52, 54, 130
Lattice gauge theories, 166
Lattice vibration problem, relation to diffusion on fractal, 134
Li,
  Li^+ ion, 161
  LiF, 161
Linear molecules, 1, 7
Limit cycle, in dendritic growth model, 88
Liquid crystal (describing hexatic phase), 54
Localization length, 147
Localized states, in disordered electron systems, 145
Marginal stability hypothesis, 73, 75, 77
Matrix oriented computers, 155
Many body perturbation theory, 155
Mean field theory, 53
  for $(GaAs)_{1-x}Ge_{2x}$ alloys, 203
  for superconducting arrays, 183
  in disordered electron systems, 149
  model for stretching of PBT, 122
Mean free path, 146
Melting, 51, 54, 101
Metal insulator transition, 51
Metal-oxide-semiconductor structures, 43
Mg, 21
Mobility edge, 147, 149
Molecular dynamics, 163
  applied to electrons on surface, 99
Molecular field theory, see mean field theory
Monolayer, 1
  monolayer-to-bilayer transition, 13
Monte Carlo,
  renormalization group, 62
  simulation of Laplacian roughening model, 60
  simulation of transitions in superconducting arrays, 177, 187
Multiprocessing, 155
Na, 21
Ne, 2, 5
Needle crystal,
  parabolic, 82

solutions, 73, 81, 85
Ni, 36
Nucleation, 76
Numerical,
  results on diffusion on fractals, 137
  results and methods in the dendritic growth problem, 83
  techniques for Hartree-Fock calculations, 161
Nonlinear dissipative systems, 74
Novaco-McTague rotation, 12
Ordered superconducting arrays, 182
Pair potential, 2
Parabolic needle crystals, 82
Participation ratio, 145, 148
Pattern selection, 85

Pd, 1, 4-5, 37
Penetration depth, 177, 179
Percolation,
  diffusion on clusters, 129, 132
  rigidity and geometrical, 125
Phase coherence length, 145
Phonon,
  mediated interactions and self-energies, 148
  spectra, 8
Photomicrolithographic deposition, 178
Physisorbed layers, 1
Pinwheel structure, 8
Pinning potential for dislocations, 11
Polaron formation, small, in disordered electron systems, 152
Poling, 110
Polybutylene terephthalate, 121
Polyethylene, 106
Polymeric glasses, 127
Polymers, phase transitions in crystalline, 105
Poly(vinyl chloride), 106
Poly(vinylidene chloride), 107
Poly(vinylidene fluoride), 107, 108
Polytetrafluoroethylene, 116
Polytetramethylene terephthalate, 121
Prokovsky-Talapov transition, 70
Proximity effect, 178, 180
Pseudopotentials, 25
Quadrupole interactions, 8
Quasiharmonic approximation, 5
Random networks, 125
Rayleigh Schrodinger perturbation theory, 159
Rectangular lattices, 10
Renormalization group, 52, 54

Registry, 1, 10, 12
Roothaan method, 159
Roughening, 51-53
Scalar computers, 155
Scaling,
  description of diffusion on percolation, 133
  of adhesive energy versus separation, 31
  of disordered electron systems, 148
  theory of relation of geometry to criticality, 130
Schottky barriers, 39
  barrier height, 39
Screening, electrostatic, 9
Self-avoiding walk, diffusion on, 133
Self-consistency, 22
Self-similarity, 130, 185
Semiconductors,
  alloy, 197
  amorphous, 149
  effective mass Hamiltonian, 165
Si, 176
  Si_xGe_{1-x} alloys, 40
Sidebranches, 80
Sierpinski gasket, 131
Sine-Gordon equation,
  discrete, 119
  double, 112
Six vertex model, 52
Sinanoglu-Pitzer-McLachlan interaction, 4
Singleton-Halsey theory of multilayer adsorption, 15
Small polaron formation, 152
Snowflake problem, 73, 83, 86
Solidification,
  dendritic, 73, 74
  front, 78
Solid-on-solid models, 51, 52, 70
Solvability selection, 85
Sparse matrices, efficient computation, 161
Spin glass, 177, 190
Substrate mediated interactions, 4
Stephan problem, 75
Stranski-Krastanov mode, 2
Steepest descent quench, 101
Sublimation curve (2D), 3
Supercomputers, 155
superconducting arrays, 177
  disordered, 190
  experiment, 193
  three dimensional, 189
Surface tension, 75
Synchrotron radiation, 122
Tailing of density of states, 152
Taylor-Chirikov map, 119

Te,
  $Hg_{1-x}Cd_xTe$, 197, 199, 202
Teflon, 116
Tetrahedron, 93
Tetratic phase, 54
Thermal diffusion field, 73
Tight binding model,
  and superconducting arrays, 184
  of disordered system, 145, 198, 203
Tip,
  parabolic, 80
  velocity and radius, 77
Topological,
  frustration, 93
  singularities, 52
Transition metals, 21, 36
Transport theory, ordinary, 145
Trilayer, 7
Two-dimensional melting, 51, 101
Two-level states in glasses, 99
Two point correlation function in fractals, 131
Uniaxial registry, 10, 15
Universality, in 'fractal sense', 134
unrestricted Hartree Fock, 156
Vector oriented computers, 155
Virtual crystal approximation, 197, 203
Voronoi construction, in curved three space, 94
Vortices, 54
Wavelength selection, 74
Weak scattering limit, in transport theory, 146
Wear, 22
Wulff construction, 69
Xe, 1, 5, 10, 13
Xy model, 180, 190
Zn, 21

Printed in the Netherlands by
Offsetdrukkerij Kanters B.V. — Alblasserdam

QC 176 .A1 M53 1984b
Midwest Solid State Theory
 Symposium (12th : 1984 :
Surfaces and disorder